Machinamenta

Machinamenta

The thousand year quest to build a creative machine

Douglas Summers Stay

Machinamenta Publishing
machinamenta.blogspot.com
dss@umd.edu

Contents

Introduction

Are machines capable of being creative? The concept at first appears to be an oxymoron—"mechanical" is an antonym of "creative." Yet for centuries, people have imagined the potential of automated thought and creativity. While they made some naïve mistakes, they often had a broader perspective on how this problem fits into the larger context of philosophy and society.

The history this book presents is a miscellaneous one. It draws from the history of the arts, magic, religion, toys, games, staged entertainment, philosophy and language. All of these threads are part of the same story: the invention of machines to automate the creation of new designs or new ideas.

The Kaleidoscope Pattern

For all their diversity, there is a pattern common to many of these devices, one that will show up again and again throughout the book. The pattern can be seen most easily in the construction of a kaleidoscope. A kaleidoscope is a very simple machine consisting of three parts:

1. Colored bits of glass.
2. Two mirrors that impose a regularity, or formal structure.
3. A means of randomizing the arrangement.

These three parts show up in virtually every attempt to make creative machines (of which the kaleidoscope is one example). They embody a theory of creativity that is centuries old: that the random rearrangement of interesting ideas or images along with an enforced logical structure is the way that our minds are able to invent new objects.

Such machines are wonderful and fascinating in their own right. One of the main purposes of this book is simply to gather many examples of these machines and exhibit them together as examples of early efforts at automating creativity. Because they fall in the cracks between art and science, many of them have been nearly forgotten by both artists and scientists, and this is a tragedy.

For all their beauty and intricacy, however, these machines are ultimately unable to deliver on their promise of true, sustained creativity. Eventually, the new images created by a kaleidoscope no longer have the power to delight and intrigue. We come to see the theme behind the variations, and each individual work no longer brings anything new to our understanding of that theme. A machine that was truly creative would be able to find ways to keep being new, and to be new in new ways.

No one has built such a machine. Despite the promise of evolutionary algorithms and machine learning techniques, every attempt so far has eventually petered out. After some initial surprises, all of these programs in the long run end up coming up only with new variations on the same themes. No programs have been run for year after year without human interference, coming up with new creations that are enthusiastically exhibited and admired.

Probably the most famous program to create visual art is Harold Cohen's AARON. AARON's work has been exhibited in major museums around the world, and its output has been described as creative by both artists and computer scientists. Each new painting created by AARON is original, and can be surprising even to Mr. Cohen. Yet he feels that despite AARON's success in the art world, it is still not creative. The reason he feels this is that AARON is essentially a kaleidoscope at heart. It has a model of a human figure, something like a paper marionette. The pose of the model, the proportions of the limbs, the placement on the page, the colors of the body segments, are all chosen randomly. The random values have constraints that keep them

Figure 1: A drawing by AARON. The pose, plant structure, colors, and shapes are randomly generated.

within reasonable bounds. A similar process generates plants and backgrounds for the scenes. Once all of this has been placed, a separate routine traces the outline of these shapes in a semi-random fashion. How is it, then, that so many people take AARON to be creative? The answer is that the program is taking advantage of certain very powerful illusions.

Illusions

A few of the machines covered in this book were intended for practical purposes, but most of them were for built for entertainment, magic, or both. These machines play on many illusions that are built into the human way of seeing the world. The early scientists drew little distinction between experiment and demonstration, and many instruments designed to illustrate a principle or entertain an audience were later used to further scientific knowledge. Showing how these illusions are built into our understanding of creative machines and processes is a second theme of the book. In the earliest case, divination machines, the devices were treated as magical by true believers in magic. Through the 1800s, automata were part of magic shows, presented as if they were magical, but with the audience aware that the magic was a carefully contrived illusion. More serious efforts at artificial intelligence beginning with invention of electronic computers inadvertently followed many of the same techniques, and had the same effect of fooling audiences into seeing the illusion of a mind, but their inventors often neglected to acknowledge the underlying illusions.

A History of Creative Devices

The title of this book, *Machinamenta*, is a Latin word that means "machines." It has only the oldest connotations—machines as siege engines, as tools of stagecraft, as ingenious contraptions. It was also used to mean clever schemes—*devices* in the other sense, or machinations. A primary meaning of *machina* in the middle ages was the cranes used by architects for building. So there is a sense of "creation" in this early definition. It was used as a metaphor in the phrase *machina mentis*, machines of the mind, to describe how the tools of memory could be used as a tool for innovation. The 17th century scholar Athanasius Kircher used *machinamenta* to describe some marvelous devices, including the self-playing Aeolian harp. So it seemed appropriate to gather under this term this diverse collection of artistic devices.

The world of computers changes incredibly quickly. Papers from a decade ago in my own field, computer vision and graphics, are almost certain to have been surpassed by more recent research that has built on them. Many of the pioneers involved with the first digital computers are still alive today. A drawback of this is that as a field, we have a very short memory. We forget that other people have been struggling with the same questions for many, many years. The problems faced in trying to build intelligent and creative machines are not merely technical, but philosophical. What is the difference between creative and derivative? What is the nature of beauty? What makes something interesting? How does the mind work?

The history of the field of computer science usually only goes back as far as World War II, with perhaps a mention of Babbage. Predictions of the *future* of the field, however, have never been in short supply. The field of artificial intelligence has more than its share of prophets, playing on the same hopes and fears that have been associated with machines that can speak to us since prehistoric times. Only by examining the project of AI in terms of its deep philosophical, mechanical, and spiritual roots can we make proper judgments about the nature of these machines now and in the future.

I
Beauty from the Symmetry of Their Form:
The Invention of the Kaleidoscope

In 1817, Sir David Brewster patented the kaleidoscope. Others had noticed the effect of two mirrors meeting at an angle before, as recounted in this selection from an 1818 article in *The Edinburgh Magazine and Literary Miscellany*:

> The repetition and reversion of images in a glass is noticed in the *Masfiti Naturalis* of Baptista Porta, a Neapolitan nobleman, who flourished about the latter part of the sixteenth century, and was distinguished for his zeal in promoting philosophical pursuits...
>
> In the *Ars Magna Lucis et Umbra* of Kircher, printed in 1646, we have an account of the same circumstance, and also of the repetition of the sectors round the centre of the circle:

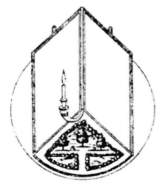

Figure 2: Kircher's adjustable mirrors

> "A wonderful property," says he, "and one which has not, as far as I know, been observed by any one, is exhibited with two specula, so constructed as to open and shut like a book; and placed on any plane in which you have described a semicircle divided into its degrees. For, if the point in which the

specula meet be placed in the centre of the semicircle, so that the side of each speculum shall stand upon the diameter, the image of an object will only be seen once, and two objects will appear, one without the specula, the true one,—and one within, the image. But if the sides be placed at an angle of 120°, you will see the image of the object within the specula twice, that is, along with the real image, three objects But if the specula intercept an angle of 90°, you will see the circle divided into four parts, and four objects; in the same manner, at an angle of 60°, you will see a hexagon with six objects.'[1]

He then applies the principle to some curious contrivances which, by his own account, filled his spectators with astonishment. With one candle he shows how to make a complete chandelier. "With angles of 120°, 72°, and 45°, you will see," says he, "with no less delight than admiration, a chandelier with three, with five, and with eight branches."

Sir David Brewster

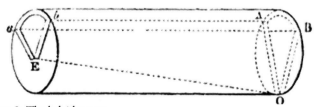

Figure 3: The kaleidoscope

Sir David Brewster, like many people of the time with an inclination to research and reading, studied theology to become a teacher and a licensed preacher. His interest in optics led to many significant discoveries about diffraction, refraction, and the use of lenses. Along with Fresnel, he was responsible for getting Fresnel lenses installed in lighthouses, and he invented the lenticular stereoscope (which uses a prism rather than mirrors to combine the stereo images). Brewster was conducting experiments on light pola-

---

[1]Athanasius Kircher, *Ars Magna Lucis et Umbra*, 1646

rization and happened on the design of the kaleidoscope. When Brewster showed his prototype kaleidoscope to manufacturers of optical instruments, pirate copies began cropping up all over the London and soon spread around the world:

> You can form no conception of the effect which the instrument excited in London; all that you have heard falls infinitely short of the reality. No book and no instrument in the memory of man ever produced such a singular effect. They are exhibited publicly on the streets for a penny, and I had the pleasure of paying this sum yesterday; these are about two feet long and a foot wide. Infants are seen carrying them in their hands, the coachmen on their boxes are busy using them, and thousands of poor people make their bread by making and selling them. (Letter from Brewster to his wife, May 1818)

The kaleidoscope allowed the viewer to enter into a virtual world, filled with bright colors and concealed symmetries. If it was a scientific instrument (as the name implied), it was an instrument of some faerie science, a science of beauty. It partook of the potential of mirrors to create other worlds, to open up new infinite spaces. The forms were reminiscent of magical mandalas, and viewers often compared the hypnotic effect of looking through a shifting kaleidoscope to that of listening to music.[2]

Brewster was an early proponent of the idea that magic and beauty could be found in technology. He wrote a series of letters to

---

[2] The idea of an analogy between color and music dated at least to 1590, when the artist Arcimboldo invented a system for composing color-music. In 1725, the Jesuit monk Louis Bertrand Castel invented an "ocular harpsichord," which opened a curtain concealing a bit of colored glass whenever a note was played. Isaac Newton was the first to realize that there may be a deeper connection in that both colors and sounds have characteristic frequencies. Despite thousands of related efforts over the years, including the light bars on an equalizer, Disney's *Fantasia*, and MTV, visual music that is able to give the same kind of effect through the eyes that music gives through the ears is still elusive.

Sir Walter Scott on the topic of natural magic. For Brewster, understanding how magic tricks and automata worked only increased
their appeal. This was in stark contrast to his contemporaries, romantic artists like John Keats, who felt that science killed beauty:

> ...Do not all charms fly
> At the mere touch of cold philosophy?
> There was an awful rainbow once in heaven:
> We know her woof, her texture; she is given
> In the dull catalogue of common things.
> Philosophy will clip an Angel's wings,
> Conquer all mysteries by rule and line,
> Empty the haunted air, and gnomed mine -
> Unweave a rainbow, as it erewhile made
> The tender-person'd Lamia melt into a shade. [3]

Brewster's conception of beauty, on the other hand, was
grounded in neoclassicism. Symmetry and geometric order were key
ideas in this. Beyond that, he assumed that a science of beauty was
possible, that universal principles of beauty could be discovered:

> If we examine the various objects of art which have
> exercised the skill and ingenuity of man, we shall find
> that they derive all their beauty from the symmetry of
> their form, and that one work of art excels another in
> proportion as it exhibits a more perfect development of
> this principle of beauty. Even the forms of animal, vege-
> table, and mineral bodies, derive their beauty from the
> same source...[4]

In *The Kaleidoscope* (a book on the optical theory behind the
construction of his invention) he gives a theory of color harmony
and repeatedly emphasizes the importance of carefully constructed
devices that don't allow the slightest imperfection in symmetry.

Yet both neoclassical and romantic concepts are evident in the
kaleidoscope: the hand selected elements are romantically beautiful,
beautiful in how they present themselves to the senses. The formal

---

[3] John Keats, *Lamia*, Part II, 1819
[4] David Brewster, *The Kaleidoscope*, chapter 20

constraints, the mirrors, are classically beautiful in how they appeal to the intellect.

It is comparatively simple to set up a system of rules and generate new images. It is much more difficult to choose a set of rules that will produce images that are aesthetically pleasing. In order to do the latter, we need to have some theory of beauty or interest. The attempt to mechanize requires that we understand; but the attempt to understand beauty transforms it. There is essentially a paradox here: creativity must continually be pushing the boundaries of what is new. Simply being new is not enough, however; to be considered creative it must be both new and beautiful. Any static conception of beauty must quickly become inadequate.

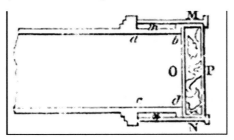

Figure 4: The end of a kaleidoscope

The problem becomes, then, how to make a machine that grows in ability over time that is not limited by the initial choices made by the inventor of the machine.

Brewster conceived of the kaleidoscope as a labor saving device for artists, an automation of part of the creative process:

> When we consider, that in this busy island thousands of individuals are wholly occupied with the composition of symmetrical designs, and that there is scarcely any profession into which these designs do not enter as a necessary part, so as to employ a portion of the time of every artist, we shall not hesitate in admitting, that an instrument must have no small degree of utility which abridges the labour of so many individuals. If we reflect further on the nature of the designs which are thus com-

posed, and on the methods which must be employed in their composition, the Kaleidoscope will assume the character of the highest class of machinery, which improves at the same time that it abridges the exertions of individuals. There are few machines, indeed, which rise higher above the operations of human skill. It will create, in a single hour, what a thousand artists could not invent in the course of a year; and while it works with such unexampled rapidity, it works also with a corresponding beauty and precision.[5]

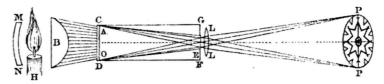

Figure 5: A projective kaleidoscope

The Reverend Leigh Richmond wrote on a similar theme:

I took up my kaleidoscope; and, as I viewed with delight the extraordinary succession of beautiful images which it presented to my sight, I was struck...with the singular phenomenon of perfect order being invariably and constantly produced out of perfect disorder—so that, as by magical influence, confusion and irregularity seemed to become the prolific parents of symmetry and beauty.

It occurred to me, that the universality of its adoption would imperceptibly lead to the cultivation of the principles of taste, elegance, and beauty, through the whole of the present and following generations, and that from the philosopher and artist down to the poorest child in the community...

I saw a vast accession to the sources of invention, in its application to the elegant arts and manufactures, and the consequent growth of a more polished and highly cultivated state of habits, manners, and refinement, in both... I was struck with the idea of infinite variety more

---

[5] *Ibia.*

strikingly demonstrated to the eye than by any former experiment. Here the sublime mingles with the beautiful.

I perceived a kind of visible music. The combination of form and colour produced harmony—their succession melody: thus, what an organ or pianoforte is to the ear, the kaleidoscope is to the eye. I was delighted with this analogy between the senses, as exercised in this interesting experiment....[6]

Things did not turn out quite as Sir Brewster and Rev. Richmond expected. Our use of machines to automate work previously done by artists has modified our concept of beauty. What used to take a great deal of skill and time could be done immediately and without effort by a mechanical process. This led to society valuing designs of rigid perfect symmetry less. A similar effect occurred with the invention of photography. Because of the ease of obtaining a perfectly accurate likeness, abstract and nonrepresentational art became more highly valued by the art world.

Brewster was at the same time able to see his invention as a toy, and as an important advance of science.[7] He saw it as a kind of proof of principle, which later engineers would use for practical purposes, in much the same way that components of automata found their way into practical machinery:

The passion for automatic exhibitions which characterized the eighteenth century gave rise to the most ingenious mechanical devices, and introduced among the higher orders of artists habits of nice and accurate execution in the formation of the most delicate pieces of machinery. The same combination of the mechanical powers which made the spider crawl, or which waved the tiny rod of the magician, contributed in future years to purposes of higher import. Those wheels and pinions,

---

[6] Rev. Richmond, *Hogg's Weekly Instructor*, Volume 6 1851

[7] At the Media Research Laboratory at NYU where I studied in 2004 and 2005, a form of kaleidoscope was actually used to study the way isotropic materials, like satin, appear and respond to light at various angles. It was a realization of Brewster's hope that the kaleidoscope would find a scientific use.

which almost eluded our senses by their minuteness, reappeared in the stupendous mechanism of our spinning-machines and our steam engines. The elements of the tumbling puppet were revived in the chronometer, which now conducts our navy through the ocean; and the shapeless wheel which directed the hand of the drawing automaton has served in the present age to guide the movements of the tambouring engine. Those mechanical wonders which in one century enriched only the conjurer who used them, contributed in another to augment the wealth of the nation; and those automatic toys which once amused the vulgar, are now employed in extending the power and promoting the civilization of our species. In whatever way, indeed, the power of genius may invent or combine, and to whatever low or even ludicrous purposes that invention or combination may be originally applied, society receives a gift which it can never lose; and though the value of the seed may not be at once recognized, and though it may lie long unproductive in the ungenial till of human knowledge, it will some time or other evolve its germ, and yield to mankind its natural and abundant harvest.[8]

Brewster was not the only one exploring the relationship between beauty and technology at the time. Mary Boole was fascinated by the mechanization of logic that her husband George and his colleagues were developing[9]. She did not have the opportunity to study mathematics herself (as she points out, women were not allowed to attend college in her day) but was eager to discuss these ideas and try to understand them in a larger context of aesthetics and religion.

Within the last generation we have gained a "Calculating-Engine," a "Calculus of Logic" (with many and widespread applications), and a "Logical Abacus;" and we are fast discovering means of making the generation of the most complicated and beautiful curves as mechanical a process as Logic has become. Of what are these

---

[8] Brewster, *Letters on Natural Magic*, 1868, p.336
[9] See Chapter VII for more about George Boole's logical calculus.

inventions a sign? The reasoning-machines of Babbage and Jevons, and the sympalmograph, and other inventions for illustrating the mathematical genesis of beauty, seem to me to have brought to a reductio ad absurdum the worship of intellectual power and artistic genius.[10]

As the machinery behind thought and artistic creation became understood, she thought, they would be valued less. The 19th century German philosopher Arthur Schopenhauer expressed a similar idea: "That arithmetic is the basest of all mental activities is proved by the fact that it is the only one that can be accomplished by a machine."

The Harmonograph

The sympalmograph or harmonograph that Mary Boole spoke of was a device for tracing out the path of combined harmonic vibrations. The pattern of Lissajous curves it traced out was a mathematically constrained image that was widely admired for its beauty. Mary Boole was especially interested in how mathematical curves could lead to the beautiful forms of leaves or flowers, a theme

---

[10] Mary Everest Boole, *Symbolical Methods of Study,*1884, p.32. Not everyone came to the same conclusion: "That art which is above all others a cultured art—that which aims at the production of symmetrical form, and the beauty which is geometrical, of man, rather than irregular, of nature, is largely a matter of machinery. For the natural and inevitable tendency of machinery is to produce symmetry.... It seems strange that many are but now awaking to the consciousness that machine-made articles need not be ugly.

But even when men have so awakened, the degrading influence of overmuch faith in machinery makes itself felt. Who can contemplate without a shudder the Corinthian cast-iron pillars of a railway station? The common mind, when it finds that some artistic process can be performed by machinery, at once jumps to the conclusion that art itself is, or can be made, a matter of machinery. This recalls the familiar story of the organ-blower, who remarked after a beautiful voluntary, 'Ah, what fine music we do make, to be sure.'" (The City of London school magazine 1877)

taken up in our century in such books as *The Algorithmic Beauty of Plants*, or *The Fractal Geometry of Nature.*

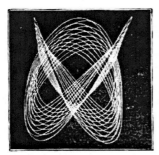

Figure 6: Sample output from a sympalmograph

Multiple versions were built, initially simple fragile oscillatory pendulums, later ones with careful arrangements of brass gears. Various inventors designed ones of the latter type which could render the same image twice with slightly offset phases, so that when viewed through a stereoscope rendered a 3D curve in space. Photographer Charles Bentham devised a way to create similar figures photographically with long exposures, writing:

It is curious to consider that these stereograms represent solid forms which never had embodied existence, and yet have had actual reality as pure form without substance, which some of the philosophers aver cannot be. The variety of the stereograms thus obtainable is infinite...[11]

It was generally judged that "successful" harmonograms required the lengths of the two pendulums to be in a ratio (such as 1:4). The chaotic curves which result from unturned lengths were judged to be less interesting and "out of harmony."

The kaleidoscope and sympalmograph are examples of a particular type of a device that is intended to generate new works of art. Each one follows the same pattern:

- Hand-selected forms to be recombined. Brewster at various points recommends buttons, bits of broken glass, a distant bonfire, dancers, coins, engravings, gems, polarized lenses, and flowers.

---

[11] Charles Bentham, "Immaterial Solids," *The Process Photogram*, Vol. 12, p. 49

- Random or nearly random input. In the kaleidoscope, this comes from shaking the bits of glass.
- Formal constraints. The kaleidoscope uses mirrors to impose symmetry.

For example, a Markov poetry generator takes a set of words (preselected for the intended effect, such as all words used in works by a given author) recombine them randomly, but imposing constraints of use frequency patterns. Fractal generators use slightly more sophisticated symmetry constraints on the randomness. Similar examples can be found for music, such as David Cope's EMI program.

Is the Kaleidoscope Creative?

When the kaleidoscope first appeared, it was hugely popular among all segments of society—rich and poor, old and young. Within a few years, however, the appeal had greatly diminished. We now see a kaleidoscope as a toy for young children. It is almost as if it were an infectious disease like chicken pox, a fascination that overtakes each of us the first time we are exposed, but we gradually become accustomed to. After a short time new kaleidoscope images fail to add anything new to the already formed impression. Instead of seeing individual works, our brains pick up on the underlying pattern that unites all the images formed. While at first it seemed that the kaleidoscope was being creative, it later becomes apparent that a store of creativity injected, as it were, during the creation of it has merely been allowed to leak out slowly.

Simple recombinative novelty, then, isn't the only thing a device needs in order to seem truly creative to us.

What I would like to do in the following chapters is explore just what it is that separates human creativity from the limited kind of pseudo-creativity exhibited by the kaleidoscope. These questions are mixed in with a tour of similar devices found in many different fields throughout history.

- How can a machine evaluate the quality of the work it produces? Can a machine be built that in some sense understands the meaning of its own output?
- How does the free will of the artist affect artwork and our perception of it? When an artist and a viewer perceive artwork, what kinds of processes occur in our minds, and can they be automated, even in principle?
- What influences our perception of beauty, art, and creativity? How can we define these in a rigorous way? Can a definition ever be provided for something that by its very nature is about discovering how to go beyond previous limits?

In the final chapter I propose a design for a machine that would incorporate a few of the more modest of these goals. Such a machine would have the ability to interpret its own products in a kind of aesthetic framework and make decisions about how to revise its output to make it more appealing.

Structure of This Book

The **first section** looks at the development of automation in various fields of art and creative endeavor.

- Chapter two looks at divination devices as randomized text generation systems. Since this is the oldest known use of this kind of machine (dating to prehistoric times), I felt it would be a good place to begin. The aspects of entertainment and illusion present in these very early systems are still an important part of how today's attempts at artificial creativity function.
- Chapter three covers devices designed to create text and poetry, showing how they follow the same kaleidoscopic principles.
- The fourth chapter is about automata, as machines that followed programs and presented the appearance of humanity.

Society responded to these machines with both horror and fascination.

- The fifth chapter explores musical instruments that compose their own music, and shows how these devices fit the same pattern that characterizes machines from all these fields.

The **second section** is on early attempts at automating aspects of the mind and brain:

- Chapter six looks at the philosophical discussion about what abilities in the mind can and cannot be reproduced in a machine, and how this impacts the ability of machines to create.
- Chapter seven is an overview of early attempts to capture rational thought and language which would eventually lead to the development of computers and expert systems.
- Chapter eight compares this approach of deductive reasoning with attempts to automate inductive reasoning and the philosophy behind it.

The **third section** focuses on 19th century inventions that led to discussions of machine creativity during that era.

- Chapter nine shows the close ties between economics and evolutionary theory in the 1800s. Scientists exploring these theories recognized that societies and ecologies could through their organization bring about creative and purposeful behavior without the direct conscious action of any individual human.
- Chapter ten looks at Babbage's computational engines from the perspective of artificial creativity.
- Chapter eleven covers the invention of photography, and how it affected the theory behind the visual arts and the role of the artist assisted by a machine.

The final chapter is a broad overview of some of the challenges in making further steps towards automating the role of the artist.

II
Machines to Generate Stories:
Board Games and Divination as
Creative Machines

Most machines have predictable output. The mill, the clock, the engine, each has a cycle that is unvarying and expected. Even in prehistoric times, however, people built a different kind of machine, devices that were generative: they produced original results not explicitly intended by their creators. One very early example is the family of divination systems used throughout Africa called geomantic systems. These are still in wide use today, and we know from inclusions in burials that they were already old when the Egyptian dynasties were just beginning. They were largely virtual machines, or software: a set of rules that if followed exactly would provide a result, rather than a physical apparatus that applied those rules.

Geomantic Divination Systems in Africa

The "hardware" of these systems is extremely simple: a grid of squares drawn in the dirt with a stick, or an array of pits dug into wood or stone, along with a handful of different colored markers. There are many variations throughout Africa and the Middle East, with layers of complexity built up over time. A typical example of their use would go something like this: the fortune-teller takes a handful of seeds and drops a few into each pit. The seeds are removed from each pit in pairs, leaving either one or two seeds in each pit. This binary code is recognized by name and used to pick out an answer to the query from a memorized structure. The code is sometimes related to the appearance of the symbol string. For example, in one system the pattern 2-1-1-1, bearing a resemblance to a flag on a flag-pole, carries a meaning of exultation (in the table below this pattern is labeled Caput Draconis, perhaps because of its resemblance to the head of the constellation Draco).

These simple patterns composed of four binary symbols are generated in groups, and the elements are recombined to derive new patterns, such as taking the first symbol from the first pattern, the second symbol from the second pattern and so forth to form a derived daughter pattern from the original mother patterns. The daughter patterns then could be recombined using addition (mod 2) to form yet another new pattern, whose meaning modifies that of the original pattern. The details are strictly passed down within a tradition, but variants exist across Africa and the Middle East. The patterns are associated not only with an interpreted meaning, but with the planets, the elements, the gods, the points of the compass, the signs of the zodiac, and so forth.

Where did such a system come from? Anthropologists can only speculate, but the same block of pits and seeds is also used for other purposes in these societies. For an illiterate population, it is a way of performing addition, subtraction, multiplication and division in a concrete way that all parties can verify, by literally reenacting the event being calculated with a single seed standing for a single item.[12] It performed functions of rewritable memory that had previously only been possible within the brain. Before the invention of writing or numbers, it was a system of symbols that represented other goods, that remapped time and space into an abstract world, with its own discrete units of space and time.

---

[12] The first abaci were drawn in the sand and used pebbles as counters, and later used pits and grooves carved in wood. The word *abacus* comes from the Hebrew *abaq*, meaning "dust." The pebbles (*calculi*) used in the Roman abacus are the origin of words such as *calculate*.

Pattern	Name	Meaning
::::	Populus	People
....	Via	Way
::: .	Tristitia	Sadness
.:::	Laetitia	Joy
:: ..	Fortuna Major	Greater Fortune
..::	Fortuna Minor	Lesser Fortune
: .:.	Acquisitio	Profit
.:.:	Amissio	Loss
.:..	Puella	Girl
..:.	Puer	Boy
.::.	Carcer	Prison
: ..:	Conjuctio	Connection
:: .:	Albus	White
: .::	Rubeus	Red
: ...	Caput Draconis	Head of the Dragon (heaven)
...:	Cauda Draconis	Tail of the Dragon (the underworld)

modified from *Games of the Gods* by Nigel Pennick, p. 55-63

Figure 7: Geomantic Divination device built in 1241 by Muhammad ibn Khuttlukh al-Mawasi.

A device which automates the steps of geomantic divination has been preserved in the British Museum. Built in 1241 in Damascus, it is a beautiful rectangular framed structure, made of brass and covered with inscribed dials, built by the metalworker Muhammad ibn Khutlukh al-Mawsili. Based on the setting of four dials to a set of binary patterns, further derived dials are set and a large rainbow-shaped area at the bottom displays the meaning of the pattern and an answer to the question being asked. The face is inscribed with the message:

> I am the revealer of secrets; in me are marvels of wisdom and strange and hidden things. But I have spread out the surface of my face out of humility, and have prepared it as a substitute for earth.... From my intricacies there comes about perception superior to books concerned with the study of the art.

From this inscription we can see that the device was personified, yet was presented without pretence as a machine. Whether using a mechanical device or simply following a set of rules, the petitioners believed that a mechanical process could behave in an intelligent way. This was their central discovery: that ideas could be held in objects, and by manipulating those ideas mechanically, one could learn something new.

Figure 9: Closeups of Fig. 1.

Another use of the same type of pits and seeds was the board game Mancala. In Mancala, the pits are said to represent fields and the markers represent seeds being sown. In this use as well, we see the board acting as a model of another activity, a simplified model with continuous space and time replaced with discrete divisions. The seeds and pits resemble the paper tape and marks along it that

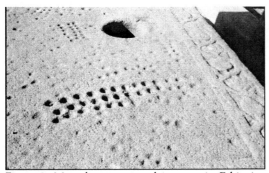

Figure 8: Mancala game carved in stone in Ethiopian archaeological site

Turing imagined in his seminal paper on computation.

The connections with modern computers are more than coincidences. The same features that made a pitted board useful for tracking heads of cattle also made it ideal for playing a game and for divination: external symbols that both players could refer to. Regarding this relationship between games and divination devices, anthropologist Wim van Binsbergen writes:

> Both material divination systems and board-games are formal systems, which can be fairly abstractly defined in terms of constituent elements and rules relatively impervious to individual alteration. Both consist in a drastic modeling of reality, to the effect that the world of everyday experience is very highly condensed, in space and in time, in the game and the divination rite. The unit of both types of events is the session, rarely extending beyond a few hours, and tied not only to the restricted space where the apparatus (e.g. a game-board, a divining board or set of tablets) is used but, more importantly, to the narrowly defined spatial configuration of the apparatus itself. Yet both the board-game and the divination rite may refer to real-life situations the size of a battle field, a country, a kingdom or the world, and extending over much greater expanses of time (a day, a week, a year, a reign, a generation, a century, or much more) than the duration of the session. In ways which create ample room for the display of cosmological and mythical elements, divination and board-games constitute a manageable miniature version of the world, where space is transformed space: bounded, restricted, parcelled up, thoroughly regulated; and where time is no longer the computer scientist's "real time" — as is clearest when divination makes pronouncements about the past and the future. Utterly magical, board-games and divination systems are space-shrinking time-machines. [13]

Considered as a way to predict the future, any existing form of divination will be little better than chance. Its interest for our purposes lies not in its accuracy, but in the way it brought people to confront the issues of artificial generation of meaning millennia before the invention of computers. As a way of holding information and allowing it to be manipulated, these techniques provided a way of working out possibilities in a safe space.

---

[13] Wim van Binsbergen, *Board-games and divination in global cultural history* (web page), 1997

Divination and Games

This same pattern played out again and again, in China, Europe, Babylonia and the Americas as well as Africa: games of chance and skill, with their discrete states and physical markers, were invariably associated with divination.

> Upon comparing the games of civilized people with those of primitive society many points of resemblance are seen to exist, with the principal difference that games occur as amusements or pastimes among civilized men, while among savage and barbarous people they are largely sacred and divinatory. This naturally suggests a sacred and divinatory origin for modern games, a theory, indeed, which finds confirmation in their traditional associations, such as the use of cards in telling fortunes.[14]

When we think of divination as a kind of game, as a way of generating new sentences from thin air, the problem of predictive accuracy is marginalized. The system was generally set up so that whatever sentence was generated would be a true sentence, because the truths encoded in the system were general truths, applying universally.

> ...The experiential (both recreational and revelatory) value of divination and board-games is that they create an unlimited variety of vicarious experiences, i.e. stories. Spinning relevant, even illuminating and redeeming stories out of the raw material which the fall of the apparatus in combination with the interpretative catalogue provides, is the essence of the diviner's skill and training; and in the same way board-games can be seen as machines to generate stories. [15]

Nearly all of the ancient board games were associated with divination at one time or another.

---

[14] Stewart Culin, *Gambling Games of the Chinese in America*, 1891

[15] Wim van Binsbergen, *ibid.*

Senet: Senet was a board game played in Egypt from around 3500 BC. Tomb paintings show the importance that Egyptian society placed on the game. A successful player of Senet was assumed to be under the protection of Ra and Thoth, since the chance fall of the throwing sticks was believed to be under their control. (For this reason Senet boards are of found among the items buried to be taken into the afterlife.)

The Royal Game of Ur: Dating to about 2600 BC, this game was played in Mesopotamia. Like Senet, it was a race game something like Backgammon. This game had certain squares thought to bring good fortune.

Go: The most prominent Chinese board game, Go was invented by the third century BC. The Go board was also used for divination, by casting the black and white stones and analyzing the patterns of how they fell. As Ban Gu described in *The Essence of Go* in the first century AD, "The board must be square and represents the laws of the earth. The lines must be straight like the divine virtues. There are black and white stones, divided like yin and yang. Their arrangement on the board is like a model of the heavens." As in Mancala, the patterns were associated with a model of the world.

Chess: There are multiple theories on the origin of chess, but one possibility is that it stems ultimately from Chinese divination methods. Chess historian Joseph Needham writes:

> The game of chess (as we know it) has been associated throughout its development with astronomical symbolism, and this was more overt in related games now long obsolete. The battle element of chess seems to have developed from a technique of divination in which it was desired to ascertain the balance of ever-contending Yin and Yang forces in the universe.... It appears that the pieces on the board in this divination technique represented the sun, moon, planets, stars, constellations, etc. The suggestion is that this "game" passed to 7th-century India, where it generated the recreational game conceived in terms of battling human armies... "Image-chess" derived in its turn from a number of divination techniques which involved the throwing of small models, symbolic of the celestial bodies, on to prepared boards. Thus there was a dice element as well as a move element, and there were many intermediate forms between pure throwing and placement followed by combat moves. All these go back to China of the Han and pre-Han times, i.e. to the -4th or -3rd century, and similar techniques have persisted down to late times in other cultures.[16]

Dr. John Dee, astrologer to Queen Elizabeth II, invented a four player chess variant called "Enochian Chess," which was designed explicitly for use in divination. Unlike random divina-

---

[16] Thoughts on The Origin of Chess by Joseph Needham, 1962

tion, it was thought that players could influence the outcome of fate through their actions on the board.

Cards: Playing card games are associated with the development of fortune-telling via Tarot cards (from which the common playing card is a simplified derivative). In the 1500s in Italy, a dealt hand of cards was used as a kind of random poetry generator. The poet would need to fit the images on the cards or their meanings into h is poem. This practice was known as "tarocchi appropriati." The fortune telling aspect of Tarot cards seems to have evolved from this game.

Divination and Mathematics

These games and divination systems are remarkably old. Consider the die used in most games of chance: the reason it has pips instead of numbers on the faces is that the form of the die settled into its present form *before* the invention of Arabic numerals.

Figure 10: Ancient Roman dice

Divination drove the development of mathematics: much of Mayan, Egyptian, and Babylonian mathematics were used for astrological purposes. For example, our measurement of time and angles come from Babylonian astrologers' division of the heavens in their base 60 system.[17] The most advanced mechanical computers from Greek and from Arab inventors in the ancient world were complex representations of the heavens, used for navigation and astrology. The Antikythera mechanism (often called the first mechanical computer) is the best

---

[17] Because 360 is a nice round number near to the number of days in the year in base 60, the ancient Babylonians divided the sky into 360 degrees. (This is easy to accomplish using a compass and straightedge.) The fact that we use 24 hour days and 60 minute hours also derive from this way of dividing up a circle.

known of these, as few others have been preserved. Found in a shipwreck and dating from around 200 BC, it showed the position of all the known planets, the sun, and the moon, requiring over 30 gears to do so. Modern scientists, who find such a device fascinating for the level of mechanical sophistication it displays, seem reluctant to admit that the only practical use such a device could have had was casting horoscopes and determining auspicious days. Watching how the planets move back and forth around the wheel of the zodiac on a recreation of this device, it is not hard to see how such an irregular motion would give the impression of an intelligent and willful plan being acted out. Early attempts by archaeologists to understand the device focused on the words inscribed on it, and were unsuccessful. It was only when an attempt was made to understand the gearing system that the meaning of the device was recovered.

Later, it was the analysis of games of chance that led to the development of probability theory and statistics, which are key components of most modern AI systems, since absolute reasoning is often too brittle to deal with real-world situations.

The combinatoric principles of the *I Ching*[18] and the geomantic divination (introduced at the beginning of this chapter)

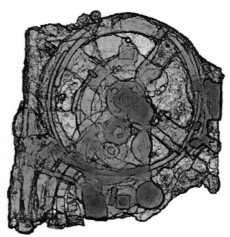

Figure 11: Antikythera mechanism, X-ray view

---

[18] The *I Ching* or Book of Changes is a Chinese method of divination that involves casting small sticks that can land in one of two possible ways. Based on the binary pattern formed by several of these casts, a fortune can be looked up in a book (thus the name).

inspired the 17th century philosopher Gottfried Leibniz to develop binary notation. These binary codes are found in other divination systems around the world, such as the African Ifa or Sikidy systems of divination. In recent years, the fields of "ethno-mathematics" and "ethno-computation" have begun studying these cultural artifacts to explore the mathematical ideas of non-Western cultures. [19]

Elements of recursion play a large role in these games and divination systems, where the state resulting from a series of actions is the beginning point for the same series of actions, performed again and again. In Mancala, for example, the game is played by choosing one pit, scooping up all the seeds from the pit, and planting one in each of the following pits. The object of the game is to be the first to get all of one's seeds into the final pit. One strategy to do this is to find a pattern that persists over time, so that the seeds in multiple pits move together in a train. These patterns were discovered and used by experienced players across Africa. In the field of cellular automata, this is known as a "glider."[20] As a form that maintains itself as it moves through a space divided into discrete cells, it is an important component in the study of these computational systems,

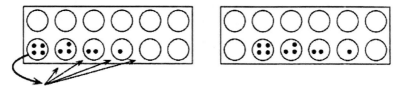

Figure 12: Glider in the game Owari, a variant of Mancala.
Illustration from *African Fractals*.

a study which only began in the 1940s as computers were invented.

These connections to mathematics are a natural extension of the representational nature of the tokens and spaces used in board games and divination. As a simpler system than the real world, it

[19] Viznut, "The Mystery of the Binary," *[Alt] Magazine*, 2003
[20] Ron Eglash, *African Fractals*, 1999

provided a fertile ground to begin development of mathematical ideas.

Divination and Ontology

The systems of classical elements (*Earth, Air, Fire,* and *Water* in Western cultures or *Wood, Fire, Earth, Metal,* and *Water* in China) used in divination rituals were attempts to find symmetries and order underlying reality, to find general systematic laws that applied to all aspects of nature and human life. These systems had appealing symmetries and provided a theoretical framework in which physics, anatomy, psychology, or any of dozens of other sciences could be understood. Most of the connections made were illusory, forced by overzealous application of symmetry, but the overall attempt to find such connections and symmetries is similar to much of modern science.[21]

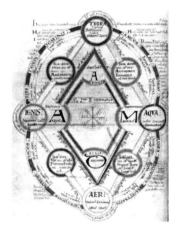

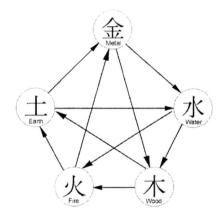

Figure 14: Four elements of the Western system.

Figure 14: Five elements of the Chinese system.

---

[21] This should serve as a warning to string theorists, for example, who also often point to the symmetrical beauty of their theory as evidence of its truth.

In his study of African divination methods, Wim van Binsbergen identified three features of geomantic divination:

- a physical apparatus serving as a random generator
- a set of rules which allow for the translation or coding of the numerical outcome of the random generator in terms of culturally agreed specific values with a divinatory meaning
- an interpretative catalogue listing such divinatory meanings and accessing them through the assigned codes

Using an assortment of pre-created elements, rules to combine them, and a randomizer, these divination systems pioneered a way of getting seemingly original creations from a machine. Are machines necessarily limited to this kind of recombination of pre-created ideas, or is it possible for them to create new works of art, new ideas which we would judge as creative if they came from a person? This is a question we will return to periodically throughout the book, as other inventors and artists used these same methods to try to build creative machines.

Cicero on Divination

The Roman scholar and philosopher Cicero examined divination critically in 45 BC in the book *On Divination*. It's hard to say exactly what Cicero believed about divination because he was careful to examine all the different possibilities in his work. One of the ideas he explored, however, was that divination might be accurate, even if it isn't guided by the gods:

> For the presages which we deduce from an examination of a victim's entrails, from thunder and lightning, from prodigies, and from the stars, are founded on the accurate observation of many centuries. Now it is certain, that a long course of careful observation, thus carefully conducted for a series of ages, usually brings with it an incredible accuracy of knowledge; and this can exist even without the inspiration of the Gods, when it has been once ascertained by constant observation what follows after each omen, and what is indicated by each prodigy.

This is remarkably similar to how digital neural networks (a form of machine learning meant to imitate the structure of the brain) are trained. At first, the correlation between input and output is completely random, but as observations are made, the associations are strengthened or weakened until it comes to accurately reflect reality in some way. Cicero imagined a simple process that would lead a system of divination to evolve over time into something that could give intelligent and predictive answers without reflecting any hidden agent providing those answers. The serious question here (one Cicero himself raises) is whether the observations were accurate enough, the correlations strong enough, and the period of adjustment long enough that the system had developed to a point where it could be useful.

Figure 15: Cicero

The Illusions of Meaning in Divination and AI[22]

Despite the fact that divination does not work, for many centuries humans nonetheless believed there was meaning in the messages generated by divination techniques. While we no longer believe in divination per se, similar illusions of meaning tend to operate in our reactions to modern generative machines. One of the main reasons we turn to AI is predictive modeling of climate, economics, or security. Divination was used for the exact same reasons. (Perhaps, following William Gibson, we could call predicting the future by means of AI *neuromancy*.) Despite divination being entirely un-

---

[22] For a careful examination of many of the cognitive issues which surround divination, see Anders Lisdorf, *The Dissemination of Divination in Roman Republican Times– A Cognitive Approach*, 2007 (PhD dissertation, University of Copenhagen).

The connection between AI and divination has been explored often in science fiction literature. *The Postman* by David Brin, for example, explores how belief shapes AI, divination, and social structures.

suited to this task (being no better than chance when done fairly, and no better than human cleverness when the system was rigged) it was widely used for millennia. That fact invites alternative explanations for its purpose.

Passing Responsibility

One possibility is that those who use divination aren't searching for accuracy but for absolution: for someone else to take over the making of decisions that are too psychologically difficult to make themselves. Attributing the decision to the fates could serve as a way to avoid criticism from others in the society. There is a strange paradox in making choices: the more evenly weighted two choices are, the more difficult it is to choose between them, but the less difference the choice makes (in the sense that the same balancing of pros and cons that makes it a difficult choice balances out the outcomes). In this case, flipping a coin is a good way to break the stalemate and take some action. Children's games, like "eenie meenie minee moe," or "rock-paper-scissors" bear similarities to divination techniques such as drawing lots, and are used primarily to make a disinterested decision. Divination could have served a similar purpose.

Entertainment

Divination was partially used for entertainment, exciting because it promised mystery and attention. (Magic 8-balls and Ouija boards are sold as children's entertainment, as modern examples.)

Figure 16: The original Ouija board.

Dispelling Worry

Just talking with someone about our dreams and worries for the future can be therapeutic. Either feeling reassured that everything will turn out all right, or being prepared for when things will inevitably go wrong, are both arguably healthier states to be in than a state of worried indecision, at least for events over which we have no control.

In addition to these reasons, there are some powerful universal illusions that contribute to our perception of such devices. Illusions come from the biases built into the brain. When such biases are applied in an inappropriate situation, we call the result an illusion. Illusions are very helpful to scientists studying perception because they give us clues to what the brain is doing behind the scenes. (Such biases are often exploited by people who want to sell you something that reason alone wouldn't convince you to buy.) Without understanding how these illusions work, it's impossible to understand why people respond in the ways they do when they interact with devices designed to imitate a mind. What ties all these illusions together is the fact that a large part of our brain is built for understanding and interacting with other people, and these modules are reused in other situations.

Illusion of Intentionality

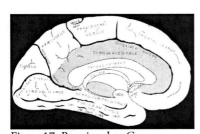

Figure 17: Paracingulate Gyrus

The perception of meaning where none is present is an extremely persistent illusion. Just as we find faces in the clouds, we are primed to recognize order so strongly that we perceive it even when it isn't present. Optical illusions are caused by the brain applying specialized modules for the early visual system in places that they are inappropriate. Divination systems were convincing because they exploited another kind of mental illusion, the mental components for recognizing intention in others.

We know quite a bit about the part of the brain used in attributing intentionality. In one experiment, people played rock-paper-scissors against a generator of random throws. Some were told they were playing against a random machine; others were told there was another player on the network. Their brain scans were compared, and the only significant difference was shown in an area called the anterior paracingulate gyrus, or PCC. People with damage to this area of the brain are unable to predict how others will behave.

This appears to be a universal human trait: we project intention and personality even when there is none present. It's inherent in how children interact with their toys, in how many religions dealt with the ideas of Fate or Fortune, in our response to dramatic performance, and in how we interact with the simple artificial intelligences in video games.

Experiments have been done since the 1940s with simple geometric figures (circles, squares, triangles) moving in relation to each other.[23] When the shapes are moved in certain simple ways, adults describe the action as one shape "helping" or "attacking" another, saying that some shape "wants" to achieve a particular goal. Infants will stare at the shapes moving in these purposeful ways longer, indicating that they are already paying more attention to things that seem to be alive.

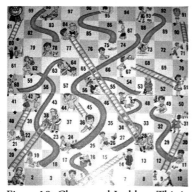

Figure 18: Chutes and Ladders. This is actually based on an ancient game involving snakes. The player makes no choices that affect the outcome of this game.

Illusion of Accuracy

Another illusion affecting our judgment is the tendency to attribute accuracy with after-the-fact judgments, known as "confirmation bias." Those predictions which happen to be

[23] Kuhlmeier, Bloom, and Wynn. "Do 5-month old infants see humans as material objects?" *Cognition*, Issue 1, November 2004, p. 95-103

true will stick out in the memory more than the others, giving an inflated impression of confidence.

Illusion of Meaning

The illusion of meaning is another link between board games, divination, and AI. Even in a game determined entirely by chance (*Chutes and Ladders*, for example, or *Candyland*), children interpret events in a game as a meaningful story, with setbacks and advantage, defeat and victory. The child is pleased at having won such a game, and feels that in some sense it shows his or her superiority. It is only after repeated exposure that, with some conscious effort, we are able to overcome this illusion. Many gamblers never do get past it, and continue to feel that their desires influence random events.

Another example is professional sports. We identify with one arbitrarily chosen set of players over another, and take their victories and defeats as our own. Yet our actions have very little influence on whether the team will be successful or not.

Illusion of Authorship

Creativity that we think is coming from a machine may actually be coming from the author of the program. The creative writing program RACTER, for example, got many of its most clever phrases directly from the programmers. In 1983, William Chamberlain and Thomas Etter published a book written by Racter called *The Policeman's Beard is Half-Constructed*, but it was never entirely clear how much of the writing was generated by the program, and how much was in the templates themselves. A sample of Racter's output:

```
More than iron, more than lead, more
than gold I need electricity.
I need it more than I need pork or let-
tuce or cucumber.
I need it for my dreams.
```

These illusions are necessary for the success of magic tricks, and for the success of computer programs that are designed to create. It may seem strange to draw such a close parallel between machines and magic. However, both words come from the same root word (the proto-Indo-European root *magh-, meaning "to be able, to have power") and have a common purpose. [24] They only differ in whether the effect is achieved by means we understand, or by means we don't. What is hidden from us is occult. Aleister Crowley wrote:

> Lo! I put forth my Will, and my Pen moveth upon the Paper, by Cause that my will mysteriously hath Power upon the Muscle of my Arm, and these do Work at a mechanical Advantage against the Inertia of the Pen …The Problem of every Act of Magick is then this: to exert a Will sufficiently powerful to cause the required Effect, through a Menstruum or Medium of Communication. By the common Understanding of the Word Magick, we however exclude such Media as are generally known and understood. [25]

With the invention of the computer, we have built the world that ancient magicians imagined already existed. It is a world formed by utterances, a textually constructed reality. The world imaged through the screen of a ray tracer doesn't resemble our world—it is instead the world that Plato described, where a single mathematically perfect Ideal sphere without location in time or space manifests through many visual spheres, which cast their flat shadows onto the pixels of the screen. The spheres are hollow: computer graphics is a carefully constructed art of illusion, presenting only on the surface.

The Turing Test

Pioneering computer scientist Alan Turing wrote a paper in 1950 exploring the possibility of whether a machine can be said to think. He proposed that a blind test, where a human asks questions in an attempt to elicit inhuman responses, would be the best way to

---

[24] Joshua Madara, *Of Magic and Machine,* 2008 (web page) The Crowley quote is also found in this essay.

[25] Binsbergen, *ibia.*

answer this question. If a human interrogator couldn't tell whether she was having a conversation with a machine or another human, the machine would pass the test and be considered to think. It remains a popular goal line that AI researchers would someday like to cross.

The point here is that the Turing Test *requires* a program to be deceitful in order to be successful. Even a genuinely intelligent machine (whatever that might mean) would still need to deceive the users into believing it was not a machine but a person in order to pass the test. The trick of getting people to believe is built into our understanding of what it means for a machine to exhibit intelligence. Turing argued that when a computer program could consistently fool people into believing it was an intelligent human, we would know that it actually was intelligent. I would argue that that threshold was passed long ago, before the invention of writing, and that we know nothing of the kind. Divination machines convinced human interrogators that there was a thinking spirit behind them thousands of years ago.

It may sound as if I am coming down harshly on AI, saying it is nothing more than a sham, merely unscientific nonsense. My intention is rather in the opposite direction: to say that meaning in AI systems comes from the same root as meaning in many of the most important areas of our lives. Like the rules we agree to when we sit down to play a game, and like language, money, law or culture, the meaning in artificially created utterances or artwork only exists to the extent that we as a society agree to behave as if it does. When we do, it can be just as real to us as those very real institutions. It can affect the world, for good or for ill, by the meaning we take it to have.

Figure 19: Alan Turing

When we speak a language, the sounds we make don't really have any significance of themselves. It is only because we all pretend that a particular series of sounds stands for a particular idea that the system of language works. If we lost faith in it, the whole system would fall apart like in the story of the tower of Babel. It's a game, and because we all know and play by the same rules, it's a fantastically useful one. The monetary system is the same way. "Let's pretend," we say, "that these pieces of paper are worth something." And because we *all* play along, the difference between playing and reality fades away. But when we lose faith in the game, when society realizes that other players have been cheating, the monetary system collapses. Artificial creativity seems much the same. If our society acts like the creative productions of a machine have artistic value, then they will have value. Value is an aspect of the socially constructed part of our reality.

In the future, more sophisticated AI systems will be better able to deal with the meaning of words, whether or not this meaning is grounded in actual conscious perception[26]. For many human purposes, though, how well an AI works is irrelevant. The way we relate to a system is largely unchanged by its accuracy or its humanness of thought. For those who want to design creative machines, this is both a blessing and a danger. We will need to think very carefully about how we design and train machines that may, someday, be better at getting their own way than we are. Norbert Weiner, the founder of cybernetics, warned about the potential of learning machines that seem able to grant our every wish:

> The final wish is that this ghost should go away.
>
> In all these stories the point is that the agencies of magic are literal minded; and if we ask for a boon from them, we must ask for what we really want and not for what we think we want. The new and real agencies of the learning machine are also literal-minded. If we program a machine for winning a war, we must think well what we

---

[26] Perceptual consciousness and the grounding of meaning are discussed in Chapter 5.

mean by winning. A learning machine must be programmed with experience... If we are to use this
experience as a guide for our procedure in a real emer-
gency, the values of winning which we have employed in
the programming games must be the same values which
we hold at heart in the actual outcome of a war. We can
fail in this only at our immediate, utter, and irretrievable
peril. We cannot expect the machine to follow us in
those prejudices and emotional compromises by which
we enable ourselves to call destruction by the name of
victory.

If we ask for victory and do not know what we mean
by it, we shall find the ghost knocking at our door. [27]

---

[27] Norbert Wiener, "On Learning and Self-Reproducing
Machines." 1961

III
Each Line Composed by This Machine:
Poetry and Natural Language Generation

When we build machines that deal with meaning, it can be hard to unravel what part of the meaning is created by each of the three participants—the machine, the machine's creator, and the one reading the output of the machine. A device for doing divination is at a disadvantage because it not only must create meaningful utterances, but

Figure 20: Christian Bök

also choose its words so that they form a valid reply to the question. There is a field of poetry called aleatory (meaning "chance") writing that uses similar techniques, but without any pretence of predicting the future. The poet Christian Bök writes, "Aleatory writing almost evokes the mystique of an oracular ceremony—but one in which the curious diviner cannot pose any queries."

Games, aleatory writing, and divination all have in common the creation of meaning. When poetry is composed with the aid of computers or randomizing elements, it raises questions about the nature and origin of meaning. Discussing such random poetry, Bök writes:

> The reader in the future might no longer judge a poem for the stateliness of its expression, but might rather judge the work for the uncanniness of its production. No longer can the reader ask: "How expressive or how persuasive is this composition?'—instead, the reader must ask: "How surprising or how disturbing is this coincidence?'
>
> ...When we throw the dice, we throw down a gauntlet in the face of chance, doing so in order to defy the tran-

scendence of any random series, thereby forcing chance itself to choose sides, either pro or con, with respect to our
fortune. Does such a challenge occur when a poet decides
to write according to an aleatory protocol? Does the poet
wager that, despite the improbable odds, a randomly composed poem is nevertheless going to be more expressive and
more suggestive than any poem composed by wilful intent?
Is meaning the stake wagered in this game? [1]

What follows are a few examples of machines designed to generate
writing or poetry through the years. Not to in any way denigrate the
cleverness of their creators, but none of them are actually very good at
writing. Even with the power of modern computers it is still impossible to generate a paragraph of sensible text on a topic without
following a very strict template. (For example, programs that take
financial data or sports scores and generate a daily news report.)

Even such simple systems, however, illustrate that meaning for a
reader in a text can be completely disconnected from the intentions
present when the text is written. Remember the story about the million monkeys typing the works of Shakespeare—a random process is
perfectly capable of creating anything that can be written, if we're
willing to put in the effort to sort through all the garbage it generates
to find the gems. This demonstrates that the key to creativity, the
really hard part, is judgment of quality, selectivity. How do we recognize good creative works when by the definition of creativity, they are
something new that we have never seen before?

The Eureka

In 1677, one John Peter published *Artificial Versifying, A New
Way to Make Latin Verses* as a kind of entertainment for schoolboys.
This explained a technique for composing Latin poetry automatically.
Each verse was of an unvarying form:

Adjective Noun Adverb Verb Noun Adjective

The meter of each word was also fixed, forcing the line into metr-
ically correct hexameter:

---

[1] Christian Bök, Harriet: Poetry Foundation Blog, "Random
Poetry", 2008 (web page)

dactyl trochee iamb molossus dactyl trochee

For example, one of the lines the machine produced could be translated as "A gloomy castle sometimes shows a bright light."

It is perhaps surprising that a method of generating poetry was invented before one for generating prose, since poetry is supposed to require more of a creative spirit. But as with divination, the most important ingredient is the license that we grant to the poet or the oracle. We have a tendency to interpret strangeness in poetry as deliberate, rather than a mistake. The foreign language, Latin, also may have served to allow an additional step of interpretation to take place in the mind of the reader, making further allowances.

Figure 21: The Eureka

In this case, the system was eventually literally automated. John Clark, an inventor and printer from Bridgewater, England, began in 1830 to build a machine to carry out the steps of John Peter's process. He built a cabinet the size of a small bookcase that composed the poem while simultaneously playing "God Save the Queen." His device consisted of six turning cylinders, one for each of the six terms in the line of poetry. If it had simply displayed six words, it would have been regarded merely as a clever plaything. But Clark encoded the words using pins in such a way that they would cause individual letters to fall into place, apparently at random. This gave the impression that the machine was somehow composing the poem *letter by letter*, which was much more impressive. He deliberately fostered this illusion, writing in a letter to the editor of *The Athenaeum*, a monthly magazine:

> Permit me, as the constructor of the Eureka, or Machine for composing Hexameter Latin Verses, to make a few observations on its general principles, in reference to Dr. Nuttall's remarks, in your last week's paper. The machine is neither more nor less than a practical illustration of the law of evolution. The process of composition is not by words already formed, but from separate letters. This fact is perfectly obvious, although some spectators may probably have mistaken the effect for the cause—the result for the principle—which is that of kaleidoscopic evolution; and as an illustration of this principle it is that the machine is interesting—a principle affording a far greater scope of extension than has hitherto been attempted. The machine contains letters in alphabetical arrangement. Out of these, through the medium of numbers, rendered tangible by being expressed by indentures on wheel work, the instrument selects such as are requisite to form the verse conceived; the components of words suited to form hexameters being alone previously calculated, the harmonious combination of which will be found to be practically interminable.— Yours, &c. J. Clark. July 2, 1845.[2]

By the time this machine was built, there was an active press in London. This makes it possible for us to follow the conversation as society responded to the introduction of a machine that could compose. Of particular interest is what verbs were used to describe the actions of the machine: it was said to be "composing," "selecting," and "thinking." The machine follows the same architecture, the kaleidoscope pattern laid out in the introduction, where individual pieces (the Latin words in this case) are randomly chosen and are combined according to strict rules. All these machines are associated with entertainment and with illusions of mental activity, in this case explicitly encouraged by the inventor.

At the time of its exhibition (for one shilling at the Egyptian Hall in Piccadilly, London) the device attracted a lot of attention. William Thackeray joked in Punch magazine that "several double-barrelled

---

[2] John Clark, *Atheneum*, 1845

Eurekas were ordered for Eton, Harrow, and Rugby."[3] One author wrote, "I do not see its immediate utility; but as something curious, it is, perhaps, entitled to take its place with Babbage's Calculating machine, and inventions of that class."[4] In fact, Charles Babbage was familiar with the machine and with its inventor. William Hodgson, an economist, wrote, "It is truly a curious machine. Though I cannot say much for the sense of the verses.... The inventor spent fifteen years upon it, five years more than are needed to make a boy into a verse-making machine, and still less perfect. Clarke is a strange, simple-looking old man. Babbage said the other day that he was as great a curiosity as his machine."[5]

On its front face was inscribed the lines:

> Eternal truths of character sublime,
> Conceived in darkness, here shall be unroll'd;
> The mystery of number and of time
> Is here displayed in characters of gold.
> Transcribe each line composed by this machine,
> Record the fleeting thoughts as they arise;
> A line once lost will ne'er again be seen;
> A thought once flown perhaps forever flies.

Part of the fascination expressed regarding many of these creative machines was the ephemeral character of their random creations, which if not recorded would be lost forever. Imagine the most beautiful poem ever crafted. Wouldn't it be an unspeakable tragedy if it was played only once, to an empty room, when not a single soul was listening? Nature is extravagant in this way with beauty. Think how many sunsets passed before there was anyone around to appreciate them, or of the clouds on Jupiter, storms the size of worlds, with no one there to watch them roll in. Or perhaps, a listener *does* hear the poem, just this single time, and forever afterwards is haunted by a few words, a single phrase. Granted, the Eureka didn't produce poetry of this caliber; but it wouldn't be hard to make a similar machine that

---

[3] Punch 9, 1845, p. 20
[4] *Littell's Living Age*, Volume VII, p. 214
[5] Life and Letters of William Ballantyne Hodgeson, 1883, p. 52

did produce such poetry occasionally, mixed in with enormous amounts of nonsense (the monkeys and Shakespeare again.) The rules that forced the Eureka to always generate grammatical sentences were perhaps too strict to allow any truly creative sentences; eliminating the possibility of embarrassing failures led to a strategy that was too conservative to allow spectacular successes. (In the last chapter there is some discussion of how some future program may be able to move past such limitations.)

Another point often mentioned in the tabloids was the combinatorial explosion of possible sentences. Over the course of a week, one journalist noted, the machine, if left running, would produce over 10,000 unique verses. It becomes a torrent, poetry enough to make a person choke. It is like a snowy waste, where the unique, delicate snowflakes pile up to form mile after mile of mind-numbing sameness.

The device also included a kaleidoscope (a fad at the time—see chapter I), which produced a unique illustration to accompany each verse. The inventor was aware that both of these machines were performing analogous functions, that what he was building was just one of a class of "creative" machines. The Eureka has been maintained and can still be seen in the Records Office of Clarks' factory in Somerset.

The idea of a poetry-generating machine was a kind of running joke from this time period until the early twentieth century. For example, "The Poetry Machine" was a short story by Charles Barnard published in 1872. In this story, a young boy happens upon a poetry machine:

> "He went up to the table and stood before the wonderful array of cranks, wheels, and levers. The machine was about three feet long and two feet wide and high. There was a clockwork attachment, with a weight that hung on a pulley under the table. It resembled a telegraph machine. There was a long ribbon of paper rolled on two wheels, and it had a marker, just as Morse's instrument has, to print the words. On one side were a number of stops or handles, with ivory heads, having curious words marked upon them. One was marked, "Serious," another, "Comic," another, "Serenades," and so on; one was marked, "Stopped

rhymes," another, "Open rhymes," and there was one marked "Metre."

The boy generates poems without meter and with nothing but the rhymes as he learns how to operate the machine. The story serves as a parody of the kind of thoughtless poetry that was churned out for commercial jingles or greeting cards. Besides similar stories, "poetic machine" was used as a humorous metaphor for the poetry-making capacity in the poet's mind (especially for poets whose primary concern was making sure each pair of lines rhymed). All of these references assume that the reader will agree that simply "turning a crank" to generate poetry is an absurdity.

The Literary Engine

In *Gulliver's Travels*, Jonathan Swift made light of devices that create language automatically and randomly. The engine was meant to parody the Royal Society, who were interested in codes and ciphers as well as the study of nature. Like the infinite shelves of books described in Borges' story "The Library of Babel," the device he describes contained all possible sentences, both sensible and nonsensical:

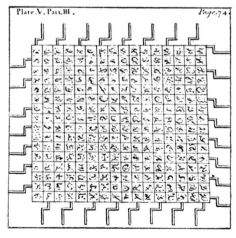

Figure 22: Illustration of the literary engine from *Gulliver's Travels*.

Every one knew how laborious the usual Method is of attaining to Arts and Sciences; whereas by his Contrivance, the most ignorant Person at a reasonable Charge, and with a little bodily Labour, may write Books in Philosophy, Poetry, Politicks, Law, Mathematicks and Theology, without the least Assistance from Genius or Study. He then led me to the Frame, about the Sides whereof all his Pupils stood in Ranks. It was twenty Foot Square, placed in the middle of the Room. The Superficies was composed of several bits of Wood, about the bigness of a Dye, but some larger than others. They were all linked together by slender Wires. These bits of Wood were covered on every Square with Paper pasted on them, and on these Papers were written all the Words of their Language, in their several Moods, Tenses, and Declensions, but without any Order. The Professor then desired me to observe, for he was going to set his Engine at Work. The Pupils at his Command took each of them hold of an Iron Handle, whereof there were fourty fixed round the Edges of the Frame, and giving them a sudden turn, the whole Disposition of the Words was entirely changed. He then commanded six and thirty of the Lads to read the several Lines softly as they appeared upon the Frame; and where they found three or four Words together that might make part of a Sentence, they dictated to the four remaining Boys who were Scribes. This Work was repeated three or four Times, and at every turn the Engine was so contrived that the Words shifted into new Places, as the Square bits of Wood moved upside down.

Six Hours a-day the young Students were employed in this Labour, and the Professor shewed me several Volumes in large Folio already collected, of broken Sentences, which he intended to piece together, and out of those rich Materials to give the World a compleat Body of all Arts and Sciences; which however might be still improved, and much expedited, if the Publick would raise a Fund for making and employing five hundred such Frames in Lagado, and oblige the Managers to contribute in common their several Collections.

He assured me, that this Invention had employed all his Thoughts from his Youth, that he had emptyed the whole Vocabulary into his Frame, and made the strictest Computation of the general Proportion there is in Books between the Numbers of Particles, Nouns, and Verbs, and other Parts of Speech.[6]

Turing's Love Letter

Perhaps the first person to use an *electronic* computer to compose original pieces of writing was Alan Turing. He wrote a love poem generator as a joke in 1947. It used random numbers to choose nouns and adjectives to fill some of the parts of a canned love letter[7]:

```
Darling Sweetheart,
You are my avid fellow feeling. My affec-
tion curiously clings to your passionate
wish. My liking yearns to your heart. You
are my wistful sympathy: my tender lik-
ing.
Yours beautifully
M.U.C.
```

Compared to the love letters of, say, Robert and Elizabeth Browning, it comes up a little short. It is similar in structure to how the utterances of fate were determined in divination rituals. The design is the kaleidoscopic pattern:

- A system of generating unpredictable output (in this case, a pseudorandom number generator).

- Elements to be recombined which themselves have some aesthetic appeal. (The words such as *sweetheart*, *wistful*, or *tender* which are randomly chosen.)

- A structure that imposes certain constraints on the results so that they will be interpreted as a sensible arrangement. (The fixed sentence grammar.)

Artists will protest that by "creative" we mean something more than this. What more could be added to a machine besides a larger number of elements and ways to be rearranged, and a more versatile and subtle structure to put them in? I think that the answer lies in the

---

[6] Swift, Gulliver's Travels III:V

[7] Lavington, 1975 from Boden, *Mind as Machine* p. 674. It would be interesting to discover who Turing intended these letters for, given what is now common knowledge about his sexual orientation.

ability to recognize beauty or quality, in order to reject generated art that doesn't meet this standard, rather than simply constraining the possible creations. Research into how to accomplish this is still in its infancy.

ELIZA

The first program designed for natural language conversation was a simulation of a psychiatrist, known as Eliza. Eliza's conversation used the patient's words in a superficial attempt to keep the conversation going on a topic without contributing much of its own. For example:

```
Well, my boyfriend made me come here.
YOUR BOYFRIEND MADE YOU COME HERE
He says I'm depressed much of the time.
I AM SORRY TO HEAR YOU ARE DEPRESSED
It's true. I am unhappy
DO YOU THINK COMING HERE WILL HELP YOU
NOT TO BE UNHAPPY
I need some help, that much seems cer-
tain.
```

Eliza is an attempt to automate (or perhaps parody) nondirective Rogerian psychotherapy, a method developed in the 1940's and 50's by psychologist Carl Rogers. Rogers recommended a technique he called "reflection," in which a therapist restates the patient's concern to show empathy and understanding, and to help the patient to find a way to solve his or her own problem. The key to this method is that the doctor is not providing solutions to the problems; the patient is providing both the problems and the solutions.

This helps us understand how meaning can be created by divination or artistic machines. When a fortune teller sits down with a client, very little actually comes from the system of manipulating signs. Instead, the fortune teller provides a way for the subconscious mind of the client to interpret a signal out of noise.

Similarly, our response to generated art is like finding shapes in the clouds or Rorschach's inkblots. The beauty and meaning come from our attempt to find something we recognize in randomness. The machine itself does not have actual experiences to draw on. But what it can do is form an effective mirror where the viewer both provides, and is affected by, the meaning.

The author of the Eliza program, Joseph Weintraub, stated in an interview:

> You can see Eliza using one basic method or, you could even say, trick: Eliza relies on the fact that the human being interprets the signals he perceives. He interprets these signals according to his needs and his interests. He projects his own image of his partner, whether this is a living human, conversing, or whether this is a living human being and a doll interacting or whatever.

He doubted the possibility of a programmed machine ever being able to actually mean the things it was saying:

> No, that's impossible. The human being becomes a human, because he is understood and treated as a human by other humans. And that's where the deepest truths come from which nourish the human being - for example trust: to trust another human. There are things, like for example a hand on your shoulder: language is closely related to this and is learnt and developed by being based on such experiences. The computer can't have these experiences.

So we return to the question of meaning.[8] Do the words in a book have meaning, after the writer has written them and before the reader has picked it up? Imagine finding a book in the library, and being moved by what is written there. If the words that one finds meaningful got there by some other process than being written by an author (say, by monkeys pounding on typewriters who got really lucky), can we say that the meaning is somehow false, not meaning at all because it wasn't *meant*? That doesn't seem right. But Weintraub's point too, seems like common sense. Meaning can't just pop up, like so many

---

[8] The study of meaning in language is called *semantics*. The phrase "semantick philosophy" was used in the 1600's and 1700s to refer to the study of divination systems. For example, in The British Apollo (Vol. III, 1708), the anonymous author writes "Bacon proposes this and several other sorts of divination as parts of rational and useful knowledge. Whatever this Semantick Philosophy was in former times..."
A better known reference is from John Spencer, *A Discourse Concerning Prodigies*, 1665.

crocuses in the spring. For there to be meaning, it seems like there has to be a mind.

It is simple to create a system that contains the fact "PARIS is the CAPITAL of FRANCE." The same system could also contain the phrase "PARIS is the CAPTAL of THE MOON," without protest. This is because it doesn't understand the words PARIS, CAPITAL, FRANCE, or THE MOON. But systems are being developed[9] that will contain the fact that capitals have to be of countries, that all countries are on the earth, that one can't have a city where no one lives, that Paris is a city, that the moon is uninhabitable, and so on— millions upon millions of facts and the logical means to connect them and derive new facts from them. Such a system would balk at being told that Paris is the capital of the moon, because it is inconsistent with the large body of facts it already contains. In this limited sense, it can be said to "understand" what the sentence means.

However, there is another sense of the word "understand" that will be discussed in Chapter VI, where we "understand" when someone mentions a particular sensory experience they have had, such as listening to music. This kind of understanding *cannot* be communicated through a network of relationships and definitions. Whenever humans understand something, it is at the lowest level grounded in this kind of direct understanding, direct experiencing, that can't be broken down further. Part of what I mean by THE MOON is what it feels like to be gazing up at it on a cold October evening. I can only point to that experience, and if you've had one similar, you can understand. If not, no amount of explaining is going to communicate it to you.

For practical purposes this doesn't make much difference, and most AI researchers are mainly concerned about practical purposes.

---

[9] Doug Lenat's CYC or the Commonsense Computing Initiative at MIT are two prominent examples, though they will probably be surpassed by other efforts soon. The Semantic Web is a related effort advocated by Tim Berners-Lee, who invented the web. "Semantic" refers to meaning; the Semantic Web is an effort to develop tools and practices that will allow this kind of automated reasoning to take place across the internet.

For artists, though, it seems to matter a great deal. For some reason, we do care whether an artist is being authentic. The idea of receiving love letters written by someone who is not actually in love, but is incapable of feeling at all and is only "going through the motions" is distasteful even if we're sure they'll keep up the pretence.

In review, then, machines made to generate text all followed a similar pattern: strict rules to guarantee grammatical correctness with a few random elements. To the extent that the text they generated was meaningful, the meaning originated in the creator of the machine or in the reader.

For all their flaws, though, these machines did generate new text every time they were run. The machines in the next chapter, while imitating many fascinating abilities, don't rise to that standard. However, they do illustrate the growing capability of machines to imitate other human faculties needed for creative expression.

IV
Tremble Into Thought:
Wind-up Toys and Artistic Mechanisms

In order for machines to act creatively, they need to act independently. The word *automaton* (Greek for self-mover) is used to refer to such machines, and what we now call "Computer Science" was originally called "Automata Theory." Automaton is also the Greek god of chance, which at first seems like a contradiction between undetermined randomness and deterministic clockwork. The connection between the two lies, perhaps, in the kinds of undetermined mechanisms mentioned in the second chapter.

Like many of the inventions discussed in this book, automata were usually toys, and were first used for entertainment, either on the stage or as the latest gadget for rich people to impress their friends. Toys allow the exploration of an idea in a safe environment. The costs of failure are lower, and because of this many inventions were first used for entertainment before being applied to serious purposes. In the history of aeronautics, for example, kites, gliders, helicopters, balloons, powered flying machines, and rockets all were widely used as toys before being put to use for human flight. Even today, the development of spaceplanes is still largely done for entertainment purposes.

The wheel was well known to ancient American civilizations but was apparently only used for toys. The development of parallel processors and graphical rendering ability in personal computers has been driven largely by the economic pressure of video games. In the same way, artificial creativity has been explored to this point mainly by tinkerers experimenting on their own, playing with the possibilities of ideas. This is partly due to the fact that the field is itself immature: no one yet knows the best way to do things. It is also seen as an art in itself—the artist finds expression through creating a generative system.

Dr. Samuel Johnson wrote, speaking of these automata, "It may sometimes happen that the greatest efforts of ingenuity have been exerted in trifles; yet the same principles and expedients may be applied to more valuable purposes, and the movements, which put into action machines of no use but to raise the wonder of ignorance, may be employed to drain fens, or manufacture metals, to assist the architect, or preserve the Sailor."[1]

Programmable Automata

Although they were designed mainly for entertainment, these machines are among the most complex that we know of from antiquity. The machines in this chapter contained many elements that later would become parts of computers, including data storage and output that varies according to the nature of the input. Some of these early automata were able to follow a program, in the sense of a series of discrete actions which were encoded in the physical configuration of part of the machine. This chapter may be seen an aside from the main theme of the book because for all their complexity, none of these machines can be described as creative. All of them, unless reconfigured by hand, would continue to produce exactly the same output no matter how many times they were run. They are important to the history of creative machines, however, because they caused people to try to understand what separated humans from machines, and having delineated those boundaries, to try to cross them.

Hero of Alexandria is the inventor credited with many of the automata we know from ancient Greece (1st century BC). A typical machine from this period is described in Hero's *Peri Automatopoietikes*. The "battery" powering the automaton was a heavy weight attached to a rope wound around an axle. As the weight descended, the axle would spin. By inserting a peg and reversing the direction of winding, the axle would spin in the opposite way. Gears attached to the axle could drive other operations, such as tripping a lever to make the arm of a puppet move in a hammering motion. In Da Vinci's notebooks are sketches of a similar system that could allow a vehicle to be programmed, by insertion of pegs into the axle, to drive forward, stop, turn, and reverse direction in a prearranged sequence. Even today, jaded as we are with robotic toys, we would find a machine

---

[1] Samuel Johnson, *The Rambler*, 1810

made of wood and rope that moves as if it knows where it is going to be a wonderful surprise.

Figure 23: Musical bird automaton

Hero also built a group of birds that would sing each in turn, when air was pumped through. This is the first known example of a sequence of notes being automatically played by a machine.

The ancient Greeks were surprisingly advanced in their constructions. The Antikythera mechanism (mentioned in chapter II) is an example of the level of sophistication their mechanisms had advanced to. Mechanical entities, like the artificial owl built for Athena by Hephaestus, were part of the universally shared myths of the time. Aristotle wrote:

> Suppose every instrument could by command or by anticipation of need execute its function on its own; …suppose that spindles could weave of their own accord, and plectra strike the strings of zithers by themselves; then craftsmen would have no need of hand work, and masters have no need of slaves. [2]

Punch cards

The techniques that were developed for this machine that could produce music or actions required the development of data storage and playback mechanisms. It was these data storage devices that would

---

[2] Aristotle, Politics, Book 1, Chapter IV

56

later inspire Babbage to design the first computer (see chapter 10). In this way, the computer can be thought of as a conceptual descendent of automata. It isn't that we are misusing a system designed to do accounting to make artistic programs: the art came first, and the math only later.

The two main data storage methods of interest are the barrel and the punched card or tape.

Figure 24: Vaucanson's duck automaton

Barrel organs were quite popular by the 18th century (see the next chapter). In 1725, Basile Bouchon, the son of an organ maker, thought to apply the data storage techniques of the organ barrel to the

Figure 25: Bouchon's loom

automation of a loom. Instead of a metal barrel with pins, he used a paper tape with punched holes. Jacques de Vaucanson (a celebrated inventor of automata, including a flute player and a digesting duck) improved this in 1745 by placing it on a ratchet driven cylinder, so that an operator was no longer needed. This caused riots among weavers afraid of losing their jobs.

The design was made robust and became widespread in 1800 when Joseph Marie Jacquard replaced the paper roll with a stack of punched cards. This allowed the design to be programmed more easily by simply replacing some of the cards. The Jacquard loom made weaving realistic images much more practical. This could reasonably be considered to be the beginning of "digital art."

Writing and Drawing Automata

The greatest builders of automata were inventors and showmen during the 1800s. At the time, automata and magic tricks were thought of as the same kind of thing. Both produced an effect of astonishment by doing what had previously been impossible by means of concealed mechanisms. Jean Pierre Robert-Houdin[3] was one of these inventors and magicians. He invented a mechanical orange tree that sprouted leaves, bloomed, and produced fruit. Later he built lifelike automata that could write calligraphy, draw pictures, and play musical instruments. Although these seemed to be acting creatively, their operation was a kind of magic trick, though no less impressive

Figure 26: drawing automaton

for that. The arm was moved by two cams that controlled the horizontal and vertical motion of each brushstroke. When the brushstroke was completed, a lever would raise the arm from the paper, the device would move on to the next cam on the shaft, and the pen would be lowered again. The arm would also move the pen over to dip into an inkwell to refill it. Similar automata had been built in the 1770s by Pierre Jaquet-Droz and his student Henri Maillardet. When it was repaired by a museum in 1928, they were unsure which of these inventors had built it. Their question was resolved, however, when the first thing the restored automaton wrote was *"Ecrit par L'Automate de Maillardet."* The information had been stored mechanically within the automaton in the form of cams since its creation.

Maillardet built a magician which could give the correct answer to question which were inscribed on metal disks. The disks had holes in

---

[3] Harry Houdini changed his name (from Ehrich Weisz) in honor of this magician. Robert-Houdin also worked on the music-composing machine called the Componium (see next chapter).

58

them, which allowed needles to pass through, altering the state of the machine so that the answer would depend on the question—an early example of portable digital information storage.

Figure 27: Drawing of dog created by an automaton

These machines are astoundingly complex and intricate, and convey subtleties of movement like simulated breathing that make them seem more real than many modern animatronics. They show what a different mindset the inventors of that time had. It would have been far easier, in a purely mechanical sense, to create a real drafting tool, something that served a practical purpose, perhaps a kind of plotter printer. But for the makers of automata, the performance of the machine was what mattered, not the end product. They wanted to make a gadget that didn't just create images, but really *drew*.

The Turk

Another of these inventors and showmen was Wolfgang von Kempelen. He created a voice synthesizer, which reproduced sounds (mostly vowels) by forming the correct shape with an artificial mouth and tongue. Later he built the Turk, which was purported to be a chess-playing automaton.

In fact, it was a trick—a chess-player actually hid inside the machinery and directed the hand of the Turk using levers. The secret of how it worked was kept for many years, and the Turk was shown all over Europe and the United States. Among those who played chess against it were Napoleon, Benjamin Franklin, and Catherine the Great.[4]

---

[4] One of the inventor/showmen who exhibited the Turk was Johann Maelzel. Maelzel stole the design of the first metronome from Diedrich Winkel. Later, when Maelzel invented a self-playing barrel-organ orchestra called the Panharmonicon, Win-

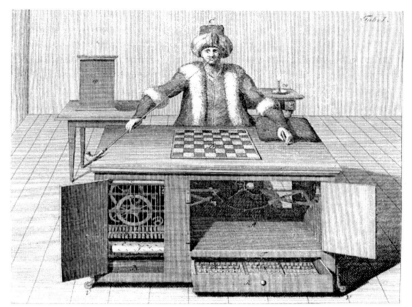

Figure 28: The Turk

The Turk provoked many discussions about the possibility of a game playing machine. Many commentators considered that each response would have to be pre-stored, and simple calculations showed that such a system would be far too large to fit within the Turk's box. The possibility of a machine acting creatively was never considered by these authors. Charles Babbage may have been inspired to think about the possibilities of a truly flexible machine by considering the Turk[5]. He discusses what would be required to build a game-playing machine in his autobiography.

---

kel got his revenge by building the Componium which, being able to compose, was clearly superior.

Maelzel seems to have had little regard for others' intellectual property. He got Beethoven to write a piece for the Panharmonicon called "Wellington's Victory," which he also stole.

[5] Tom Standage, "Monster in a Box", *Wired* 2003

Edgar Allen Poe also analyzed the Turk. He wrote that if it were actually a pure machine, it would be even more impressive than Babbage's (unbuilt) engine, because no matter how wonderful, mathematical operations are determined and unchanging, while there is no fixed response to a chess move. In other words, a creative machine would be inherently more impressive than merely a calculating one. The remainder of his article puzzles out how the trick is performed, in a manner very similar to how he would later write the first detective stories.

The author E.T.A. Hoffmann, believing that the Turk was a machine, wrote that "it serves as an oracle," making the connection between game playing and divination traced out in the second chapter. The life we see in the action of the machine, he believed, stems from a projection from our own minds. "In a dream," he wrote, "a strange voice tells us things we did not know.[6]"

The Turk affected others in a different way. The Reverend Edmund Cartwright, on witnessing the Turk and purportedly believing it to be a true automaton, decided that building a power loom would be simple by comparison, and went home and invented one.[7]

This is perhaps the main contribution of automata to the history of creative machines. By forcing the viewer to confront the question of humans as machines (and vice versa), they inspired thinkers to consider what it was, exactly, that separated people from machines and forced many authors to consider how far the imitation of human abilities could be pushed.

---

[6] Hoffmann, "Automata," 1814. Perhaps Hoffman was thinking of another automaton, the Euphonia. This was also dressed as a Turk, but unlike the chess-player could speak through the action of a mechanical trachea, larynx, and jaw. The Euphonia was played like a musical instrument to generate speech sounds in several languages, even singing the national anthem, "in a hoarse sepulchral voice...as if from the depths of a tomb." See also Chapter IV, regarding the uncanny.

[7] http://www.cottontimes.co.uk/cartwright02.htm

V

A Soft Floating Witchery of Sound:
Self-Playing Musical Instruments and
Automatic Composition

Music is a mystery. It affects our emotions directly, in ways that we can't quite put into words. It seems to be deeply tied up with the spiritual aspect of life. The idea that music can be created without a musician evokes a surprisingly wide range of reactions, from fascination to anger and fear. In some ways, though, music lends itself to automatic generation. The notes are discrete, and often constrained by well-understood rules of harmony. The generation of appealing new melodies is not yet as well understood, but by rearranging phrases of other melodies, something new can be created.

Barrel Instruments

Figure 29: Musa brothers' musical automata

In the 9th century, the Musa brothers in Baghdad invented a self-playing flute. A rotating drum with pins (similar to the drum on a music box) tripped levers that uncovered holes on the flute. The drum was powered by a water wheel and water flowed into an air-filled chamber to force air out through the flute. This drum was an early example of a machine controlled by a reconfigurable binary data storage device (though there may well have been such machines earlier which we have no record of). This data storage mechanism is invaluable to music researchers because it provides us an accurate reproduction of exactly what the original listeners would have heard, unlike early musical notation which is often more or less obscure.

The idea behind the flute was never entirely forgotten in Europe, and organs built on the same principle show up by the 1300s. Barrel organs were originally full size instruments, but were soon miniaturized into boxes that could be carried or carted around by street musicians. Organ grinders were eventually a common sight in many European cities.

The satirist Cyrano de Bergerac imagined that the music box could be made to play voices as well as musical notes, in an early work of science fiction called *The Other World*, in which a Candide-like character travels to the moon. Other scientists at the time were inventing physical ways of generating speech sounds, so this is not as absurd as it seems:

> When I opened a box, I found inside something made of metal, somewhat like our clocks, full of an endless number of little springs and tiny machines. It was indeed a book, but it was a miraculous one that had no pages or printed letters. It was a book to be read not with eyes but with ears. When anyone wants to read, he winds up the machine with a large number of keys of all kinds. Then he turns the indicator to the chapter he wants to listen to. As though from the mouth of a person or a musical instrument come all the distinct and different sounds that the upper-class Moon-beings use in their language.
>
> When I thought about this marvelous way of making books, I was no longer surprised that the young people of that country know more at the age of sixteen or eighteen than the greybeards of our world. They can read as soon as they can talk and are never at a loss for reading material. In their rooms, on walks, in town, during voyages, on foot or on horseback, they can have thirty books in their pockets or hanging on the pommels of their saddles. They need on-

ly wind a spring to hear one or more chapters or a whole book, if they wish. Thus you always have with you all the great men, both living and dead, who speak to you in their own voices.[1]

Barrel organs, music boxes, and the like were capable of playing music, but certainly not composing it. However, some of the same inventors working with these systems were working on the problem of automated composition as well.

Automatic Composition

The simplest methods of generating note sequences without a musician all used wind for power. Wind chimes and wind bells from Southeast Asia date back to at least 3000 BC. To introduce further randomization, wind chimes usually hang from a string. The weight of the chime on the string naturally swings back and forth with a regular beat. But the force of the wind hits this cycle with varying strength at different points in the swing. Such a "forced pendulum" exhibits chaotic, unpredictable behavior, even if the wind is constant in strength. Similarly, the design of the Aeolian harp introduces eddies into the wind that flow chaotically over the other strings, playing them randomly. The study of chaos and turbulence has only begun in the last 40 years or so, but these instruments have been making practical use of it for centuries.

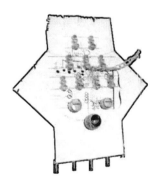

Figure 30: Musical kite

Kites with bamboo whistles attached were invented around 700 AD in China. The practice was so common that the modern Chinese word for kite, *fengzheng*, taken literally means "wind zheng," after the musical instrument called a *zheng*. Each whistle generates a single tone, varying in intensity depending on the speed of the

[1]Cyrano de Bergerac, *The Other World: The Societies and Governments of the Moon* Chapter 34, 1659

64

air flowing through it, but typically a kite will have multiple whistles attached and many such kites may be flown at once. Drums, gongs, bowed strings and other instruments were also attached to add strings and a percussion section to the aerial symphony.

Athanasius Kircher and the Aeolian Harp

Figure 31: Aeolian Harp

Athanasius Kircher probably invented the Aeolian harp. "Probably" because a lot of things he claimed to have invented we know were copied from others; but in this case, his claim seems legitimate. Kircher was a Jesuit scholar who wrote on an astonishing variety of scientific and religious subjects in the 17th century. The books are eclectic, intriguing, occasionally occult. Perhaps the most comparable figure who is well known to the general public today would be Leonardo DaVinci. Kircher's books were enormously popular and respected in his day, but as the first real experimental scientists (such as the Royal Society in England) began to test his pronouncements, they found he was wrong as often as right. His interpretations of hieroglyphics, for instance, are comically bad. His reputation was damaged and he was largely forgotten. But he shows up again and again in the history of artificial creativity. Through him we know of many of the ancient automata, methods of automatic composition, the first idea of the kaleidoscope, systems for automatic translation, and many other areas of applied technology that didn't seem worthy of study by later (more theoretical and less whimsical) scientists.

Kircher belonged to the tradition of *natural magic*, which over-lapped more or less with science from the 17th through the 19th centuries. Perhaps the closest thing to natural magic today would be a science museum show: the goal was to astound and to amaze by the use of hidden devices, in a way that would lead the viewers to curiosity about the explanation. This was in contrast to natural philosophy, which descended from the philosophical tradition which had no place for experiment and was more concerned with geometric and logical demonstrations. We tend to think of natural philosophy gradually becoming science as it began to make use of experiment, but it was only by adopting the methods, technology, and tools invented for natural magic that it was able to do so. Newton's experiments with prisms, for example, were only possible because he was able to buy the prisms which were already being used for natural magic.

Kircher called the Aeolian harp *"Machinamentum X"* and *"Machina Harmonicam Automatam."*[2] It consists of several strings stretched over a sounding board. The harp is placed where wind can blow over it, often in front of a window that has been left open a crack. The eddies introduced into the wind by the first string cause the others to vibrate. The name "Aeolian" comes from the story of king Aeolus in the Aeneid, who trapped the winds in the caves of a mountain and forced them to do his bidding.

The Aeolian harp became popular in England and Germany with the rise of the Romantic Movement in the late 18th century. It has

---

[2] It is likely others had noticed the effect earlier: in a Midrash, for example, King David is said to have been awakened by his harp sounding when the north wind blew across it in the morning. The self-playing harp in the fairy tale "Jack and the Beanstalk" may also stem from a similar cause.

been called "The Romantics' scientific instrument."[3] The Romantics saw the wind harp as a symbol of creative inspiration by nature. For example, it is featured in "Ode to the West Wind" by Shelley, and "Dejection: an Ode," and "The Aeolian Harp," by Coleridge. Coleridge made explicit the analogy between mental operations and the blowing of the wind:

> And many idle flitting phantasies,
> Traverse my indolent and passive brain,
> As wild and various, as the random gales
> That swell and flutter on this subject Lute!
> And what if all of animated nature
> Be but organic Harps diversly fram'd
> That tremble into thought as o'er them sweeps
> Plastic and vast, one intellectual breeze,
> At once the soul of each and God of all?[4]

The sound of the wind harp is something ghostly and alien, and has a tendency to make the hair on the back of your neck stand on end. Coleridge called it "a soft floating witchery of sound." They were associated with inspiration by ghosts and spirits. Aeolian harps illustrate artificially creative machines greatest strength and greatest weakness—the property of being literally inhuman, inconceivable within the usual ideas of what art should be.

The Uncanny

This eeriness was mentioned by E.T.A. Hoffmann in a short story entitled "Automata." (Freud later referred to Hoffmann's stories in his famous essay "The Uncanny.") While he praises the sound of the harp, the main character, Lewis, finds other music created by machines awful: "The gravest reproach you can make to a musician is that he plays without expression; because by so doing he is marring the whole essence of the matter. Yet the coldest and most unfeeling executant will always be far in advance of the most perfect machines."

---

[3] Thomas Hankins and Robert Silverman, *Instruments and the Imagination*, 1999, p. 112.
[4] Samuel Taylor Coleridge, "The Aeolian Harp," 1796

He also says that machine generated music is "tantamount to a declaration of war against the spiritual element in music." Perhaps it was this unease that led to Joseph Gabler, inventor of the "Vox Humana" organ stop, to have been accused of having come up with the idea through a deal with the devil.

This uncomfortable feeling is also associated in the story with the appearance of machines that try to imitate humans:

> "All figures of this sort," said Lewis, "which can scarcely be said to counterfeit humanity so much as to travesty it— mere images of living death or inanimate life—are most distasteful to me. When I was a little boy, I ran away from a waxwork exhibition I was taken to, and even to this day I can never enter a place of the sort without a horrible, eerie shuddering feeling…. It is the oppressive sense of being in the presence of something unnatural and gruesome; and what I detest most of all is the mechanical imitation of human motions."

Cognitive scientist Masahiro Mori pointed out in the 1970s that this problem happens whenever something approaches humans either in appearance or motion. We respond to a puppet or animated character better the more lifelike it is, but only to a point. After that, the closer resemblance to a human is also closer resemblance to a moving corpse, and we are repulsed. Presumably, if a simulation were exact enough, we would pass this point and respond favorably again, when we became unable to distinguish between the simulation and reality on a subconscious level. He dubbed this positive-negative-positive response curve the "Uncanny Valley." While it isn't really a scientific theory—there isn't one dial labeled "realism" we can dial up or down, and see what effect it has on people—it is a well known effect among computer graphics artists, animators, and researchers working to recreate human abilities.

Arca Musurgica

Athanasius Kircher also created a system for the mechanical composition of music he called the *Arca Musurgica*. It consisted in a stack of cards with a series of notes on each card, which could be rearranged according to a set of rules to form a new melody and associated harmonies in counterpoint.

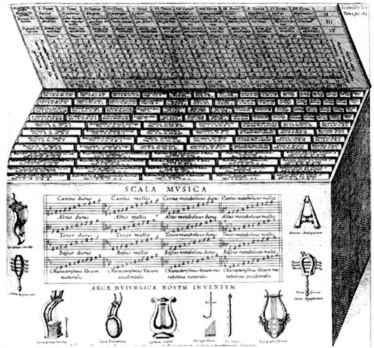

Figure 32: Arca Musurgica

This is similar to the kaleidoscope in that it takes small pieces which themselves have some beauty and rearranges them randomly but with an imposed symmetry to ensure that the final composition is pleasing overall.

The Arca Musurgica was just one set among many similar sets of tables covering a dozen subjects. Kircher assembled everything he knew how to calculate into tables on small cards: cards on Arithmetic, Geometry, Fortifications, Ecclesiastical Calendars, Sundials, Astronomy, Astrology, and Cryptography. He kept all these in a portable case he called the *Organum Mathematicum*. (Kircher tutored the children of nobles, and presented the *Organum* to one of his students, Archduke Karl Joseph of Habsburg.) In an era where computation was a

Figure 33: Water driven barrel flute

difficult manual process, a set of pre-computed tables could perform many of the functions we would today rely on a computer for. The box also included a gadget to aid in multiplication based on the principle of Napier's bones.

Kircher's system of music composition was hardly the first system of its kind. Perhaps the first person to write down a set of rules for composing music was Guido d' Arezzo, who invented our modern musical notation around 1000 AD. He set up a correspondence between vowel sounds in lyrics and musical notes. The same vowel sound would always be sung to one of a small set of possible notes. The names for notes we use today (*Do, Re, Mi, Fa,* etc…) come from his identification of the notes and sounds of the hymn *Ut queant laxis.*

Guillaume de Machaut was a fourteenth century composer who created short sequences of notes and rests that were repeated at different tempos to create original compositions. In the fifteenth century, canons were composed by specifying a rule (or "canon") for how the different voices should come in a certain number of measures later.

70

Musical Dice Games

Figure 34: Mozart's dice music

Mozart later invented his own system for composing music by means of dice. Each measure was chosen randomly (by the sum of the two thrown dice) from a list of measures appropriate for that point in the composition. For a few of the measures the selection is very limited, ensuring that the final composition will sound acceptable. Johann Sebastian Bach also composed using algorithms, and some of his well-known pieces were first specified as a set of rules, and only later were written down using traditional notation.

The Componium

In 1821 the Dutch inventor Diedrich Winkel[5] created the first device which combined automatic composition with the automatic playing of the barrel organ. His Componium was a cabinet-sized device that played back music from two cylinders. While one was playing, the other moved randomly to a different position and played a different variation (from among eight possibilities) on the next set of measures. It was able to do this by means of a pulley which, when loosened, spun freely like a roulette wheel until it reengaged, determining whether the selector moved left, right, or stayed in the same line on the barrel. About this invention one author wrote:

> Already mechanical science has succeeded in binding down the wings of genius; and carpet manufacturers and fancy workers no longer consult the taste of artists, but apply to the kaleidoscope to supply them with new patterns. Are musical composers in future to be taught to take their inspiration from such an instrument as is now exhibiting? [6]

Winkel also built automated looms and invented a way to produce innumerable weaving patterns, though little record of this exists. When the Componium fell into disrepair, the great magician and inventor of automata Jean-Eugene Robert-Houdin set about repairing it. It took him a full year and by the end he was so mentally exhausted it took him five years and a trip to France to recover.

The idea that a machine could compose music was nothing short of a revelation to some of those who went to see it. The French educator Jean-Joseph Jacotot wrote about the Componium as a metaphor for the subconscious mind. He had been asked to teach French to a class of Flemish speaking students (a language he himself did not speak) without any teaching materials. Under his enthusiastic guidance, as a class they approached the language as a puzzle, noticing regularities, grammatical rules, and cognates, and began to form hypotheses about the meaning of sentences. Everything they discovered they wrote down and recited daily, until they had formed a respectable

---

[5] See footnote 5 from Chapter 2.
[6] The Times, 20 May 1830, from *Componium* by Van Tiggelen p. 89

dictionary and grammar for the language. In the mean time, through the process of discovery, the class had learned much more quickly and deeply than comparable students in traditional classrooms, who were taught at the time mainly through rote memorization.

Figure 35: The Componium, showing the composition barrels

Jacotot's teaching method became very popular in France, and he wrote several books explaining how the principles he had discovered might be applied to teaching many different subjects. In his book on music, he describes the Componium (representing genius, or the creative mind) as a powerful natural force, overcoming all obstacles:[7]

> The Componium is a machine.

> Genius is an instinct.

---

[7] J. Jacotot, *Enseignement Universel. Musique*, 1824. In the spirit of his teaching method, I have attempted to translate the passage despite never having learned French (though with the help of automatic translation algorithms.)

If we believe everything that has been published, if reports are true, the Componium reproduces a given theme in any number of different forms— new, varied, and wonderful. It is not exhausted; the thousand variations it has produced do not impair its fecundity. It is always full.

The genius never fails.

At a height at which the eye can't reach, a source unknown and hidden to our weak eyes, this overflowing torrent rushes; it surprises, it frightens, it takes everything that is easy and that is difficult and still gives an encore. Rolling with its rushing waves over all the obstacles he has encountered, he is used to defeating his opponent; his power, his incessantly accrued conquests, fall with all their weight on those in attendance. He still walks with them, he runs faster, he feels less resistance, has fewer stops, fewer detours, the less he stops his rapid growth. The improvisation begins. Listen, you cannot guess the order that follows, he does not know himself. An infinite number of roads lead to the goal, but of all the roads that have been taken, it turns away to a new course. His march is planned in advance but he ignores it. It is taken and it takes us, it is captivated when it charms us, and it shows he believes a higher being does reveal him to himself.

More than any other area of artistic expression, musicians have been successful in developing algorithms for creating interesting, even beautiful compositions.

The story of David Cope and his Experiments in Musical Intelligence (EMI) is perhaps the greatest success story in artificial creativity. As computers able to work with musical notes first became available, Cope began writing software to help him with his own composition process. By 1987, he was using the software to create new compositions in the style of earlier composers.

Cope coded the entire scores of several pieces by a particular artist in a format his program could read. The program would then take small snippets of these scores and recombine them, rearranging according to fixed rules to make new compositions.

The structure of the new pieces resembled the large scale structure of the original scores because of these fixed rules. It was the design of these rules, based on a deep understanding of musical structure, that causes the scores to resolve in plausible ways. On the small scale, the use of exact musical phrases made it appear to have been written by the same composer.

EMI was profoundly successful. Cope would visit music schools, and ask the audience to vote on which was the human and which the machine composition. The results were stunning: these professional musicians were unable to recognize which had been human composed. But the reaction among musicians was mostly negative. They found the whole idea devalued human creativity. When he tried to get professionals to play EMI's compositions, he was repeatedly rejected.

Douglas Hofstadter, a philosopher who has written heavily on machine creativity, was deeply shaken by his experience with EMI. It cast into doubt his conviction that there was a profound human soul that he could sense behind classical music.

In 2003, Cope destroyed the training data, leaving only EMI's own compositions to learn from. The new software, trained on these compositions, is called "Emily Howell."

Is Emily Howell an example of true machine creativity? I would argue that one essential part is still missing. Not all the compositions that EMI creates are at this professional level; some contain clear mistakes. Cope actually chooses which compositions to keep and present to the public. It seems that a human selection of quality is still a necessary part of the creative process.

This issue doesn't seem insurmountable in this case, though; a few more heuristics may be enough to eliminate any real blunders. A deeper issue is that the compositions based on a particular composer all have a similar flavor. In overall structure, the approach Cope used still follows the kaliedoscope pattern. Like any kaleidoscope, it doesn't take too many observations of the generated patterns before the differences between them begin to seem unimportant and the similarities become more and more obvious.

Is there a way to overcome this problem? Or are there fundamental differences between our minds and computers that can never be overcome? This question is ultimately a philosophical one, and the next chapter takes us deep into the thicket of philosophy of the mind.

VI
Crystals of the Mind:
The Philosophical Limitations of Machines

In essence immaterial, are these minds,
as it were thinking machines?

For, to understand may but rightly be to
use a mechanism all possess,

So that in reading or hearing of another,
a man shall seem unto himself

To be recollecting images or arguments,
native and congenial to his mind:

And yet, what shall we say,—who can
read the riddle?

The brain may be clockwork, and mind
its spring, mechanism quickened by a spirit...

Doth normal Art imprison, in its works,
spirit translated into substance,

So that the statue, the picture, or the
poem, are crystals of the mind?

Tupper's Complete Poetical Works

Even the first philosophers faced the question of the relationship between mind and mechanism. Epicurus suggested that everything, including thoughts and perceptions, could be explained by the motion of atoms. Others believed in a dualistic philosophy. The question was still completely open through the 1700s. For example, in 1697, in an editorial in The Athenian Oracle, John Dunton listed these aspects of the mind as impossible to reproduce mechanically:

free will : "We have a Root of Liberty, which nothing of Matter can pretend to."

perception: "Perception is not mere reaction of Matter, but a recognition of those Impressions which have been formerly made, a Flight much too high for Matter. Nay, the Body is a perfect Statue or Machine, without the actual Operation and Advertance of the Soul."

abstract thoughts : "[The mind] can form abstracted Notions, and even strip Matter of itself in Demonstrations, and Mathematical Universals."

conscious attention: "…if we do not attend to it, if the Mind does not fix itself on the Object immediately before it, but ranges and wanders somewhere else, we are still never the wiser; unless it starts and, as it were, shakes itself into reflection, tis not conscious of those outward Actions, it knows not what we read, or see, or hear."

memory: "It seems inconceivable that the prodigious number of Ideas ranged in the Memory should be corporeal; if they were, where would there be room for them, or how could they but confound one another, as an infinite number of pictures in a Glass would do? Much less is the calling forth of any of them in such admirable Order, a work of Chance or Matter; any more than a thousand Alphabets shook together, and then exposed to a Looking-glass, could by virtue of the Glass immediately throw themselves into a Poem, or an Oration."[1]

Dunton went on to say that, "It appears as incongruous to talk of Rational Matter, as of a Yard of Sound, or of the Colour of a Thought."

However, mechanical ways to realize a few of these aspects of the mind were eventually found. Mechanical calculators worked operations on numbers and returned an abstract result. With the work of Charles Babbage[2], the analogy between mechanical memory and human memory also became accepted. These devices were slow and large, but the process of miniaturization had already begun (pocket watches, for example, had become practical by the 17th century) and the possibility of some of these aspects of the mind being performed by mechanical means in the brain grew in popularity. In time, only the mechanical reproducibility of the first two, subjective perception and free will, would be left as open questions.

---

[1] Dunton, *The Athenian Oracle*, Vol. III, p. 368
[2] See Chapter X for more on Babbage.

Subjective Sensations or Qualia

Take a deep breath and let it out slowly. Feel the air come in through your nose, smell the odor of the room, feel your chest expand and rub against the fabric of your shirt. Compare this to a breath taken in deep, dreamless sleep: when one is unconscious, breathing still occurs and is regulated by the nervous system, but is not accompanied by any of these sensations. It happens "automatically'—that is, in the manner of automata.

Our bodies can react to certain stimuli even when completely unconscious. For example, the pupils of unconscious patients will contract when a bright light is shining into them. This is not accompanied by any sensation of brightness.

Philosophers refer to these internal, subjective sensations as "qualia." Is it possible, even in principle, for a machine to experience qualia? When humans and inanimate objects were considered to be radically different kinds of things, it was easy to simply deny that such a thing was possible. But as more and more aspects of the mind were reproduced mechanically, the possibility that qualia, too, could be reproduced began to seem more of a possibility.

One way out of the problem is to claim there is something different about brains (for example, some kind of connection to a non-physical world), and that they are able to generate qualia in some way that copies in another medium that behave identically are not able to do. While this may seem appealing, it has some drawbacks as a scientific theory.

Presumably, there is some way that the activity of the brain influences this non-physical realm, and is in turn influenced by it. Suppose that in your friend this activity were rerouted, so that instead of creating and being influenced by qualia, the neural signals are sent into a computer that calculates what the influence of qualia would be, and sends those signals back. When you talk to your friend, he insists that nothing has changed. "I feel fine," he says. "When I look at green objects, I still have the same sensation of green I always did." Your friend would have to respond this way; the silicon brain is functionally identical to the original. Yet no matter how much he insists, to maintain our position that brains are special we would have to accept that with the silicon brain there are no qualia present, that he is acting entirely unconsciously.

The other way out is to allow the astonishing idea that *all* physical systems have qualia. Leibniz went ahead and took this philosophical step. He started from the following premises: (a) people have subjective perceptions, (b) their bodies are a kind of machine, and (c) that "in any mill or clock taken by itself no perceiving principle is found that is produced in the thing itself; and it does not matter whether solids, fluids, or a compound of both are considered in the machine." From this he argued that the ability for perception of a simple kind must exist without the machinery of the brain. This led him to conclude that all things have a perceptive part. This also provided him with a neat explanation of how matter could react to gravity and other distant forces—it perceived all the surrounding matter. This is the core of his theory of monads:

> It must be confessed that perception and that which depends upon it are inexplicable on mechanical grounds, that is to say, by means of figures and motions. And supposing there were a machine, so constructed as to think, feel, and have perception, it might be conceived as increased in size, while keeping the same proportions, so that one might go into it as into a mill. That being so, we should, on examining its interior, find only parts which work one upon another, and never anything by which to explain a perception. Thus it is in a simple substance, and not in a compound or in a machine, that perception must be sought for.[3]

Leibniz's idea is that all particles have a little bit of consciousness. He argues it is in the particles themselves, and not in their arrangement that consciousness comes about. The idea that the *arrangement* could be responsible for consciousness is a currently popular philosophy called computationalism. Computationalists believe that whenever the functional aspects of the brain associated with qualia are reproduced in a computer, the qualia will also be present. Despite this

---

[3] Gottfried Leibniz, *Monadology,* Section 17 (Robert Latta translation)

distinction, the two philosophies end up having very similar consequences.

It's not too hard to imagine that a robot like C-3PO could be conscious. The trouble with this line of thinking is that like Leibniz, we have to go all the way. Since we can call any change of states a "computation" after the fact, this seems to imply that in the air and in the rocks, perceptions of exquisite pains and subtle pleasures are popping into existence without any reason. (This is technically referred to as the "dancing pixies" problem.) The implication is even worse than that: the exact same set of state changes could be mapped to the execution of a program purportedly causing the sensation of the color green and mapped to the execution of another which is supposed to evoke the color red. Which is the correct interpretation? Computations are subject to external interpretation, but qualia are not.

In the third chapter, there were simple computing devices made of cogs and gears, or hydraulic systems, or ropes and pulleys. Calling the interaction of these parts a computation is a matter of perspective, an interpretation we place on the system from the outside. We could similarly call the interaction of molecules in any liquid a computation. What the inputs, outputs, and state changes are can be assigned after the fact, as we choose to interpret the collisions. [4] In order to maintain that a particular computation implies consciousness, it seems necessary to add the caveat that the consciousness is only there, only real, in its own context. As one commenter writes, "Like fairies are real in the medium of fairy stories? That's what we call 'not real' when we're being serious."[5]

---

[4] In the same way, any random combination of letters can be associated with some varying cipher that allows it to be transformed into any particular English sentence of the same length. The qualia of the computation in the first case seem to be no more inherent in the computation than the meaning of the sentence in the second case.

[5] Peter Hankins (channeling John Searle) on consciousentities.com

Free Will

The will is another aspect of the mind that may not in principle be automatable. Most professional philosophers and AI researchers believe that free will is compatible with determinism—that the future of the world is entirely caused by the present state of the world, but that somehow we are still free to choose without being constrained. If that is true, then the will could in principle be implemented in a machine. Certainly, something indistinguishable in *behavior* from a will could be built into a machine. Partial determinism with the addition of random chance is the basis of most of the machines discussed in this book. Some philosophers like Robert Kane and Galen Strawson believe otherwise, however. It is hard to see how moral responsibility could be assigned to a machine. If a machine today injured a person, we would hold the builders of the machine responsible, or call it an accident. How is that supposed to change as the machines grow more complex and subtle? Their movement would still be completely determined by their program.

A related issue is strength of will. Most of our heroic literature is written in praise of willpower, the ability to stick to a resolved course of action even when pain makes doing so difficult. Think about the experience of making a hard choice: it can be literally painful as the pull of the two sides force themselves onto our attention and we need to resist one side or the other. Now imagine a program making the same kind of choice. It may have to spend more computational resources on the decision, but is there anything we could find to praise in its coming to the right decision? The machine would simply do whatever fell out of the computations.

Choices in a machine are either arbitrary, or determined, or some combination of the two. This doesn't seem the same as bravely choosing, for example, to express an idea in the face of ridicule or attack.

The creation of a work of art also involves numerous choices. These can be hard choices, requiring mental effort. Part of what we find worthy of praise in an artistic creation is the work that went into creating it, the hours of labor in becoming the kind of person who could create such a thing. Would we find anything similar to praise in a machine artist? Or would all the praise go rightly to the programmer?

Ghost Stories

All of these difficulties tend to reinforce the widely held notion that there is something deeply mysterious, even super-natural (in the sense of being outside of natural, physical explanations) about the mind. This kind of thinking is completely unacceptable to most scientists.[6] It takes the thing we are most interested in understanding and put it into a phantom realm completely off limits to scientific inquiry. Instead of explaining qualia or choice, it is postulating a homunculus (tiny person) outside the physical world (but at the same time attached to the brain) that does the perceiving and pulls the strings to control the body.

---

6 John Locke, the 17th century English philosopher, pointed out that just because we can't imagine how a mechanism could think, doesn't make it impossible:

"The objection to this is, I cannot conceive how matter should think. What is the consequence? Ergo, God cannot give it a power to think. Let this stand for a good reason, and then proceed in other cases by the same. You cannot conceive how matter can attract matter at any distance, much less at the distance of 1,000,000 miles; ergo, God cannot give it such a power: you cannot conceive how matter should feel, or move itself, or affect an immaterial being, or be moved by it; ergo, God cannot give it such powers: which is in effect to deny gravity, and the revolution of the planets about the sun; to make brutes mere machines, without sense or spontaneous motion; and to allow man neither sense nor voluntary motion.

Let us apply this rule one degree farther. You cannot conceive how an extended solid substance should think, therefore God cannot make it think: can you conceive how your own soul, or any substance, thinks? You find indeed that you do think, and so do I: but I want to be told how the action of thinking is performed: this, I confess, is beyond my conception: and I would be glad any one, who conceives it, would explain it to me...By the same reason it is plain, that neither of them can move itself: now I would ask, why Omnipotency cannot give to either of these substances, which are equally in a state of perfect inactivity, the same power that it can give to the other?" John Locke, *An Essay Concerning Human Understanding*, Vol. II, 1768, p. 146

In a machine this whole operation could be replaced with a computational process. One could take the weights which would be assigned to qualia, partially randomize them, and compute output choices. If the distribution of choices coming out of the module is indistinguishable from the distribution coming from a person, the two would be functionally similar even though internally they are completely different. If this is true, it would be possible to make a machine which acts just like a human, but has no interior mental life. (The technical term used by philosophers for this is "zombie." Philosophers of the mind have an odd sense of humor.)

If qualia and free will are *not* a form of computation, then what could they possibly be? Secretions of brain chemicals? Something quantum mechanical? Some kind of oscillation? Or something outside the natural world entirely? We don't know, and it may be literally impossible for science to find out. In order to build a machine that we are certain experiences qualia, we would have to have a test that unambiguously detects when qualia are present or absent. Because qualia are subjective, such a test seems to be, by definition, impossible. Perhaps the best we can do is understand everything else about the brain, so that we can understand qualia by the shape of the hole, so to speak.

Free Will, Qualia, and Machine Creativity

This all bears on the problem of machine creativity. When a human paints, the action is accompanied by sensations of pleasure, happiness, frustration and desire, by the qualia of colors and the smell of the paint. A major motivation for creating art comes from the pleasurable qualia in the artist, that the artist anticipates in the audience, and that the artist gets from people paying attention to the work.

So when we say a computer can be a creative artist, we have to be careful. A program may generate works that we judge to be interesting or beautiful or creative. There is, however, (at least so far) little reason to believe that the experience of creating the artwork will be anything like what human artists go through as they create art. If the creation of art isn't a struggle—against laziness or the limitations of perception and skill—we tend to value the creation itself less. No one has yet proposed any kind of system that could create a greater work of art by "really pushing itself" or "trying harder."

None of this is really new. In Medieval Europe, the soul was thought, for religious reasons, to be peculiarly human. And yet animals were observed reacting to their own senses, forming plans, and remembering past events. Without a soul, these functions would have to be performed by mechanical means in those creatures.

Perhaps the main point of disagreement between philosophers Rene Descartes and Thomas Hobbes was on this issue. Descartes took the position that the human brain was somehow associated with a separate spiritual consciousness (the "I" in the phrase, "I think, therefore I am." His argument is actually more along the lines of "I perceive, therefore I am.") Hobbes, on the other hand, would have stood with the modern computationalists, arguing that consciousness was a byproduct of the machinelike operations of the brain.

The passage below, by the French philosopher Paul Janet[7] writing in the mid 1800s, attempts to show that animals (as opposed to people) act in a manner completely constrained by mechanical laws. He was echoing arguments made by Descartes.

> ...we have seen that the operations of instinct themselves differ in nothing essential from the functional operations of the living machine... If, then, a simple agency of physical causes, without any foresight, express or implicit, can explain how living nature succeeds in accomplishing the series of delicate and complicated operations which terminate in the structure of an organ, why should not the same mechanical agencies produce a freak, no doubt more complicated, but not essentially different — that of an animal that has the air of feeling, thinking, and willing, without possessing any of these faculties?...

---

[7] Paul Janet doesn't get much respect among scholars: The 1911 Encyclopedia Britannica, for example, gives him just two paragraphs and writes of his works, "They are not characterized by much originality of thought." Like Gottfried Leibniz and Athanasius Kircher, two other philosophers whose ideas are explored in this book, he is just starting to get more attention as the topics he addressed become more relevant to our everyday lives with the increasing importance of computers.

I say, then, that mechanism cannot urge any serious objection against the automatism of the beasts; but the same mechanism ought to go much farther still, and ought not to recoil even from the automatism of men — I mean automatism in the strict sense, namely, a mechanism purely material, without intelligence, passion, or will. If the animal is only a machine, why should other men be anything for us but machines? ...After all, what proof have we that other men are intelligent like ourselves? ...we only know ourselves immediately — we have never directly discovered intelligence in other men. It is only, then, by induction... that we assume in other men a mind and an intelligence as well as in ourselves. ...Now, if a combination of causes can have produced, without any art, what so closely resembles art, why could it not have produced, equally without any intelligence, what would as closely resemble intelligence? The hypothesis is not so absurd, since there really are cases in which men act automatically and unconsciously, as if they really were intelligent — for instance, cases of somnambulism or of dementia.... see whether it is impossible to refer to chance the formation of an organism resembling ours so as to be taken for it, manifesting entirely similar action, but which would only be a fiction — an automaton in which not a single phenomenon could be discovered having an end, and which would consequently be destitute of all intelligence....[8]

Here he is proposing that a purely mechanical system could behave like an intelligent being. The term "intelligent" seems a little out of place here. If a machine is able to answer the questions correctly on an intelligence test, is able to act in a way that would, in a human or animal, be considered intelligent, why would we deny the adjective to the machine? Janet is referring to something else when he uses the

---

[8] Paul Janet, *Final Causes,* 1876. On this topic Leibniz wrote: "There is no doubt that a man could make a machine capable of walking for some time through a town, and of correctly turning at the corners of certain streets. . . . Those who show the Cartesians that their way of proving that the brutes are automata would justify him who should say that all other men except himself are simple automata also, have justly and precisely said what I mean." *Riplique aux reflexions de Bayle : Opera philosophies,* pp. 183, 184, ed. Erdmann

term. He denies that we can know that other people are also intelligent, but we are, in a practical sense, able to determine fairly easily whether a person we meet is intelligent or a fool. The issue that we are unable to resolve by observation of others is whether or not such intelligent or foolish behavior is accompanied by the same kind of internal perceptions and actions of will that we experience ourselves (the solipsist's dilemma). Such confusion is still fairly common in the philosophical literature dealing with AI: for example, the much discussed "Chinese Room" argument attempts to show that statements generated in a mechanical way are not "understood" by the mechanical process. Demonstrating "understanding," in the sense of being able to correctly answer questions about Paris and the Moon that was addressed earlier, is an automatable process. The real questions are whether such "understanding" behavior is accompanied by the appropriate qualia or guided by free will.

If an animal can act with intentions, why not the larger system we call nature? Whatever "intentions" are in animals, according to Janet, there is nothing that prevents the same kind of thing from showing up in a disembodied system, such as the evolutionary process itself:

> If I have the right to suppose that [an] animal pursues an end when it combines the means of self-preservation and of self-defense, why might I not suppose, with the same right, that living nature has also pursued an end, when, as wise as the animal, she has prepared for it the organs which are for it the fittest to attain that end? ...

> I do not know whether the mechanical philosophy has ever taken account of the difficulty of this problem... Is it not evident that for a brain to think, it behooves to be organized in the wisest manner, and that the more complicated this organization, the more probable is it that the result of the combinations of matter will be disordered and consequently unfit for thought?

> Thought, in whatever manner explained, is an order, a system, a regular and harmonious combination ... That these innumerable applications might become possible, it has been necessary that millions of living and sentient cells, only obeying, like printers' types, physical and chemical laws, without any relation or resemblance to what we call intellect, should be assembled in such an order that not on-

ly the Iliad, but all the miracles of the human intellect should become possible. For if these cells, in their blind dance, had taken some other direction, some other motion...not reason, but madness, as experience shows, would have been the result; for it is known that the least blow given to the equilibrium of the brain suffices to undo its springs and arrest its play.

He points out that there must be within the living body some kind of in-built tendency towards order, since any kind of disorder results in madness rather than reason. [9] While Paul Janet wants to keep the idea of some kind of god directing things, he thinks of that god as nature whose goals are brought about through evolution.

We know nothing, absolutely nothing, of the cerebral mechanism which presides over the development of thought, nor of the play of that mechanism. But what we know for certain is that that mechanism must be extremely complicated, or, at least, that if it is simple, it can only be a wise simplicity, the result of profound art. Whether this very art be the act of an intelligence similar to that, the mystery of which we are investigating, we will not now inquire here...

It is impossible to dissemble the blunt intervention of chance in this evolution of natural phenomena, which, hitherto governed by the blind laws of physics and of chemistry, the laws of gravity, of electricity, of affinities...is suddenly coordinated into thoughts, reasonings, poems, systems, inventions, and scientific discoveries.

...this would be yet once more be a true miracle, and a miracle without an author, that thought should suddenly originate from what is not thought. In order to diminish the horror of such a prodigy, it will be supposed that the molecules of which organized beings are composed are perhaps themselves endued with a dull sensibility, and are

---

[9] The idea of a law of nature driving towards increased order was out of favor for many years, but has begun to resurface in research into self-organized critical systems. At first glance it seems to violate the second law of thermodynamics, the tendency of any system to degrade into disorder. The reason it doesn't violate the second law (known as entropy) is that the systems under discussion are very far from equilibrium, being driven by energy that ultimately comes from the sun.

capable, as Leibnitz believed, of certain obscure perceptions, of which the sensibility of living beings is only the
growth and development. I shall answer that this hypothesis, besides being entirely gratuitous and conjectural, after
all grants more than we ask; for, sensation being only the
first degree of thought, to say that all things are endued
with sensation is to say that all is, to a certain extent, endued with thought. "All is full of God," said Thales. All
nature becomes living and sensible. Neither sensation nor
thought is any longer the result of mechanism.

Many of Janet's ideas were derived from the works of German
philosopher Georg Hegel. Hegel proposed that consciousness pro-
gressed in each individual through various stages: sensory
consciousness, perceptual consciousness, understanding consciousness,
and eventually self consciousness. He saw this same evolution of consciousness in society as a whole, calling it "The We that is an I."

According to Hegel, logical thought starts with a thesis, or a belief. Gradually internal contradictions in the thesis become apparent,
leading to an antithesis. The attempt to reconcile these contradictions
leads to something new, a creative thought that hadn't previously been
recognized: a synthesis. This was a theory of how new ideas could be
generated from old ones, and the same process could be seen playing
out within an individual mind and in society as a whole. It was almost
as if society was a kind of universal mind that held concepts and
through conflict generated new ideas. In the same way that the mind
realizes its freedom of will, societies themselves can become freer by
becoming aware of themselves and examining their fundamental
structures of power.

Philosophers' understanding seems to have come full circle. The
earliest philosophy was one of animism, where everything contained a
spirit, and the spiritual world and the physical world were one and the
same. The number of things held to have a spirit dwindled over time,
until by the Middle Ages it was commonly taught that only humans
had spirits, and that the spirit and body were completely separate. As
science grew in explanatory power, the separate existence of consciousness from the human brain seemed less and less likely to
educated people. This chain of reasoning led to a surprising result,

however: if everything could be explained by atoms and forces, then our own consciousness must be explainable by the same. If atoms and forces could cause subjective perceptions in us, then why not in other animals? Why not in other systems that were not animals, like computational devices? Furthermore, if in computational devices, which can be instantiated with gears, water pipes, ropes and pulleys, groups of people or silicon, why not in other arrangements of gears, pipes, ropes, societies or silicon? Why not in the motion of atoms in the air, or the currents of a stream? The result is a philosophy that in a way is a return to animism, in that everything around us is held to have at least the potential for a mind.

Unfortunately, all this work hasn't led us to an answer. At the beginning of the chapter we asked whether there are there any functions of the mind that can't be reproduced by a machine. Two possibilities presented themselves: qualia and free will. Both seem to serve important roles in human creativity, and the question of whether it is possible even in principle to replicate them in a machine is still open.

Consider the following thought experiment. A piece of artwork is released on the web. It becomes very popular, and critics praise it highly as being interesting, beautiful, and creative. Later it is revealed that the artist was a machine without any feelings, any consciousness, or any will. There are three popular responses to this thought experiment:

1. What you are describing is impossible. No such artwork could exist.
2. The critics were wrong; the work appeared creative and good, but with the additional knowledge of how it was created, we can now realize that it was nothing special.
3. Art without intention is possible.

The first response will eventually be proved wrong. Remember the story of Shakespeare and the monkeys—it is simply a fact that any possible arrangement of symbols can come about by a random process eventually.

The difference between the second two responses is a difference in understanding of the word "art." As such, no creation by such a machine could ever convince those who respond in the second way that creativity in machines is possible. For such critics, the lack of perception or will is an automatic disqualification from creativity.

That's what we deserve for trying to get answers from philosophy, I guess. The next chapter leaves these boggy fields and looks at an idea that occupied many of the best minds throughout the dark ages and the renaissance: can we use the tools of language to automatically bring about the processes of thought? I attempt to show how the results of such efforts were crucial in motivating those who invented the first computers.

VII
Logic of the Future:
Leibniz and the Perfect Language

> My invention contains all the functions of reason: it is a
> judge for controversies; an interpreter of notions; a scale for
> weighing probabilities; a compass which guides us through
> the ocean of experience; an inventory of things; a table of
> thoughts; a microscope for scrutinizing things close at
> hand; a telescope for discerning distant things; a general
> calculus; an innocent magic; a non-chimerical cabala; a
> writing which everyone can read in his own language; and
> finally a language which can be learnt in a few weeks, tra-
> velling swiftly across the world, carrying the true religion
> with it, wherever it goes.
>
> – Gottfried Leibniz[1]

Ramon Llull (1232 to 1316) was a Franciscan missionary, scholar,
linguist and philosopher. He attempted to convince Muslims of theological propositions as a form of missionary work. Because he saw
truth as one great whole, he believed that if he could get his listener to
agree to any of certain propositions, he could convince him through
pure logic of the rest of the truth. In order to do this in a form he
could share with others, he decided to create a great work (*Ars Magna*)
that would establish all possible arguments on a topic from all possible
starting points to all possible conclusions.

The *Ars Magna* is a method and a machine for inventing ideas and
arguments that can be used to convince others of the truth. It is a way
of generating new arguments automatically, without effort. This, and
its influence on later thinkers such as Gottfried Leibniz, gives it an
important place in the beginnings of artificial intelligence.

Many scholars believe that this system was itself inspired by Arabic divination devices similar in spirit to those discussed in the second
chapter. Using the position of the sun and the moon (and some com-

---

[1] From Paolo Rossi, *Logic and the Art of Memory*, 2000, p. 191

plicated rules that served to add some pseudorandomness to the
process) an astrolabe with arrows on the surface would point to certain
Arabic letters, which stood for words that formed the answer to any
posed question. Other devices left out the connections with astrology
and used other randomization techniques. One specific form of the
latter type, called the "za'iraja," has specifically been associated with
Llull's *Ars Magna*.

Figure 36: From the Ars Magna of Ramon Llull

This combinatory work included a series of tables, lists of beginning attributes, and three concentric spinning disks: the first known
use of paper engineering (i.e. pop-up books). The arguments themselves are very medieval in their topics of interest. For example, one
might prove that "Goodness is in harmony with what is glorious, what
is glorious is great, therefore goodness is in harmony with what is
great." Writer and scholar Jorge Borges points out that

> in mere lucid reality... [Llull's machine is not] capable
> of thinking a single thought, however rudimentary or falla-
> cious... For us, that fact is of secondary importance. The
> perpetual motion machines depicted in sketches that confer
> their mystery upon the pages of the most effusive encyclo-

pedias don't work either, nor do the metaphysical and theological theories that customarily declare who we are and what manner of thing the world is. Their public and well known futility does not diminish their interest. [2]

Llull's efforts at persuasion must not have been completely successful; he died at the hands of the Saracens in 1316. The problem of combinatory explosion in possibilities is one of the major challenges that face AI to this day.

Throughout the medieval period, there was intense interest among scholars in the idea of creating a logical order that would categorize the world. Reference to this cluster of ideas can be found in the works of such authors as Cicero, Petrarch, Hobbes, Descartes, Francis Bacon and Athanasius Kircher under the name of the *Ars Memoria*. Dividing up the world in this way was seen as a way to discover the laws of nature, and at the same time to consolidate them into an order that could be easily memorized. It is difficult for us today to understand the importance that memorization played before books and other forms of artificial memory became ubiquitous. Developing a system of the world was necessary to enable the facts about the world to be stored compactly enough that they were possible to memorize. The connection between data compression and scientific understanding was recognized by thinkers throughout this period. They typically assumed that the categories they were discovering were identical with the fundamental reality of nature.

This tradition culminated in the work of Leibniz, who recognized that while Llulls's choice of terms was capricious and his execution of the idea simplistic, the idea of a machine for performing reasoning by mechanical combinatoric means was not, in itself, unreasonable. Llull had referred to his work as an *Ars Inveniendi*, a systematic method of invention. Leibniz was to adopt the same terminology. The invention they were interested in was invention of proofs—automatically discovering a mathematically valid proof (or disproof) of any concept, not just in mathematics, but in any field of philosophy, politics, or other human endeavor. Such a device, if it existed, would be an invaluable

[2] Jorge Borges, "Ramon Lull's Thinking Machine," *Selected Non-Fictions*, p. 155

oracle. This continued the work that had been proposed by Pappus of Alexandria when he invented the idea of "heuristics," meaning the science of invention. Beginning at age 12, Leibniz started work on this project which was always the closest to his heart, though it went uncompleted at his death.

The project consisted of five goals, any one of which would be ambitious:

1. An ontology which contains all concepts expressible by language.

Figure 37: Leibniz built this mechanical calculator, one the the first ever invented

An ontology can be thought of as a kind of dictionary, in which all ideas are defined by simpler ideas, until we reach the simplest concepts, the alphabet of thought. It was a classification of the entire world, similar to Linnaeus's classification of life into kingdoms, phyla, and so forth, but more ambitious in scope. The Dewey decimal system is another well known example of the top levels of an ontology, though more arbitrary in organization than what Leibniz had in mind.

Leibniz also drew inspiration from the ontology John Wilkins was developing in England. The part of this project that dealt with physical quantities was developed into the metric system of measurement. Like the metric system, the rest of the project also tried to replace traditional language and culture to impose a new logically simple arrangement. The project was an ambitious one, and the most complete realization of a universal ontology by that time. However, like all such projects, it suffered from preconceptions and locking into one particular worldview. As the scholar Umberto Eco writes, "In reality, the image of the universe that Wilkins proposed was the one designed by the Oxonian culture of his time. Wilkins never seriously wondered whether other cultures might have organized the world after a different fashion, even though his universal character was designed for the whole of humanity."[3]

Postmodern theorist Michel Foucalt wrote about the arbitrariness of taxonomy, in the introduction to his 1966 work *The Order of Things*:

> This book first arose out of a passage in Borges, out of the laughter that shattered, as I read the passage, all the familiar landmarks of my thought —our thought, the thought that bears the stamp of our age and our geography— breaking up all the ordered surfaces and all the planes with which we are accustomed to tame the wild profusion of existing things, and continuing long afterwards to disturb and threaten with collapse our age-old distinction between the Same and the Other. This passage quotes a 'certain Chinese encyclopedia' in which it is written that 'animals are divided into: (a) belonging to the Emperor, (b) embalmed, (c) tame, (d) sucking pigs, (e) sirens, (f) fabulous, (g) stray dogs, (h) included in the present classification, (i) frenzied, (j) innumerable, (k) drawn with a very fine camelhair brush, (l) et cetera, (m) having just broken the water pitcher, (n) that from a long way off look like flies." In the wonderment of this taxonomy, the thing we apprehend in one great leap, the thing that, by means of the fable, is demonstrated as the exotic charm of another system of thought, is the limitation of our own, the stark impossibility of thinking that.[4]

While Borges' Chinese encyclopedia is a fictional exaggeration, similar examples can be found in the study of other languages. In Japanese there is a much more extensive system of counters than in English, where we have a few special categories like "loaves" to count

---

[3] Umberto Eco, *The Search for the Perfect Language*, 1997 p. 239
[4] Michel Foucalt *The Order of Things*, 1966

bread and "pairs" to count pants or glasses. Some of the more unusual Japanese counters are:

chō	挺	Guns, sticks of ink, palanquins, rickshaws, violins
hai	杯	Cups and glasses of drink, spoonfuls, cuttlefish, octopuses, crabs, squid, abalone, boats
chō	丁	Tools, scissors, saws, trousers, pistols, cakes of tofu, town blocks
hon	本	Long thin objects, rivers, roads, ties, pencils, bottles, guitars, telephone calls, movies
ki	基	Graves, wreaths, CPUs, reactors, elevators, dams
men	面	Mirrors, boards for board games, levels of computer games, walls, tennis courts
wa	羽	Birds, rabbits

These are not classes of objects that any Western culture would come up with.

An ontology is at the heart of the Semantic Web project, and is the basis for common sense expert system projects such as CYC. The very highest levels of the CYC ontology are interesting from a philosophical point of view, since they define what are usually taken to be core concepts that can't be broken down further in terms of still simpler ideas. For example, in this system, physical objects are treated as a particular type of event: they have a beginning, an aging process, and an end. The atoms themselves continue on, but their existence as an object is a matter of coming together for a time, like a group of people attending a football game becomes the object known as a crowd. We call a football game an "event" and a football an "object" but the two can be seen as examples of the same kind of process.

Leibniz considered a clever way of transforming an ontological tree into a spoken language. Each time the tree split into branches, the branches could be named with consonants in alphabetical order. The levels of the tree would be indicated by successive vowels. From this a word such as *badefobu* would describe the path to take through the tree in order to find the meaning of the word. Once one had memorized the tree, the meaning of any unfamiliar word would be immediately apparent from its pronunciation.

2. A simplification and sharpening of grammar

Adverbs would be reduced to adjectives (*slowly* would become *slow*), which in turn would become nouns (*slow* would become *slowness*). Verbs would also be reduced to their gerund form, which is a noun. (For example, in the sentence "Running is my favorite activity," the subject of the sentence is the word "running," the gerund form of the verb "run.") Gender, declension, and so forth would be eliminated. With every part of the sentence reduced to a simple noun, operations designed to work on nouns could be applied to the meaning of the entire sentence.

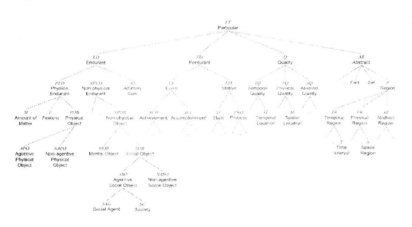

Figure 38: The root levels of DOLCE, an ontology for medical expert systems

3. A set of ideograms representing every simple concept

This *Characteristica Universalis*, as he called it, would be the perfect language. The meaning of unknown compound Latin words such as *kaleidoscope* or *telephone* can be inferred from knowledge of the root words for beauty (*kaliedo*), vision (*scope*), distance (*tele*), and sound (*phone*). Leibniz sought to invent a new language where the meaning of *any* word could similarly be guessed from a small number of roots. These roots were to be written by means of a small drawing illustrating the meaning, like the icons on a modern desktop computer. He

98

was aware that the Chinese writing system worked along somewhat similar principles. [5]

What set Leibniz's approach apart from his contemporaries was his plan to make understanding a mechanical process. Each root word could be assigned a prime number, and the primes would multiply together to form a unique large number associated with any possible word.

Figure 39: A medallion Leibniz created showing the binary numbers from 0 to 15

To find out whether a telephone is used over a long distance, one could divide the number of the word *telephone* by the number of the root *tele*. If the answer has no remainder, the answer is true. If not, the answer is false. The word in this language for man, he suggested, would be formed by combining the roots for "thinking" and "animal."

---

[5] "Such a nomenclature, in which the name of each thing (or idea) would be an adequate and transparent symbol for it and, as it were, its description or logical portrait, would clearly constitute a sort of natural language, such as Plato dreamed of in the Cratylus. It would be the Adamic language, as it was called by mystics, that is, the nomenclature that, according to Hebraic legend, the first man established in the terrestrial paradise and which men spoke until the confusion of languages at the Tower of Babel ... Leibniz thought that this supposedly primitive language was certainly unknown to us. Hermann von der Hardt asked him if the Adamic language was not Hebrew. Leibniz replied, "Saying that the Hebrew language is primordial is the same as saying that the trunks of trees are primordial"; and he added that the only question is to know whether Hebrew is closer than the others to their common root, otherwise unknown, and that this would be the work of comparative philology." (The *Logic of Leibniz* by Louis Couturat, Chapter 3 and footnote)
Leibniz believed that the study of language was the clearest way of forming an understanding of the workings of the mind.

This invention is crucially important (though terribly impractical as described). When we map the set of words to the set of numbers, we gain the ability to use all the tools that have been developed for automatic manipulation of numbers for the automatic manipulation of words. Leibniz wrote:

> The greatest remedy for the mind consists in the possibility of discovering a small set of thoughts from which an infinity of other thoughts might issue in order, in the same way as from a small set of numbers all other numbers may be derived.

4. Logical operations to act on these symbols: an algebra for language.

Leibniz believed that the processes of thought occurred by means of signs or representations on the mind. "If there were no characters," he wrote, "we could neither think of anything distinctly nor reason about it."[6] He approved of Thomas Hobbes' notion that all human thought was essentially computation.[7]

Figure 40: George Boole

> This [binary] calculus could be implemented by a machine (without wheels in the following manner, easily to be sure and without effort. A container shall be provided with holes in such a way that they can be opened and closed. They are to be open at those places that correspond to a 1 and remain closed at those that correspond to a 0. Through the opened gates small cubes or marbles are to fall into

---

[6] *Dialogue*, 1677; G VII, 191 (A&G, 271)

[7] On the other hand, he didn't believe that thinking was all there was to the human mind. He wrote, "There is a true unity which corresponds to what is called the I in us; such a thing could not occur in artificial machines, nor in the simple mass of matter, however organized it may be." (Liebniz, *New System of Nature* 1695)

tracks, through the others nothing. It [the gate array] is to be shifted from column to column as required.[8]

This idea was carried forward by George Boole and eventually became the basis of the logical operators we use today for mathematics, philosophy, and AI. Boole introduces his system:

> Let us represent by a letter, as y, all things to which the description "good" is applicable, i.e. "all good things," or the class "good things." Let it further be agreed, that by the combination xy shall be represented the class of things to which the names or descriptions represented by x and y are simultaneously applicable. Thus, if x alone stands for "white things," and y for "sheep," let xy stand for "white sheep." (*An Investigation of the Laws of Thought*, 1854)

In addition to these rules, Boole let the number 0 stand for the empty set and the number 1 stand for the universe as a whole. The class represented by xx would simply be the same as the class x. (Using the definition of x above, this would mean something like "white things that are white things.") This can be represented as the equation

$$x = xx$$

which can be rewritten

$$x = x^2$$

and subtracting from both sides

$$x - x^2 = 0$$

which is then factored into

$$x(1 - x) = 0$$

This can be understood as demonstrating the truth of the following statement: the class of things which at the same time belong to a particular category x and *don't* belong to that same particular category (*1-x*, or the set of everything, *1*, with x taken away) is empty. Boole also demonstrated the use of such algebra in syllogistic reasoning. Leibniz seems to have anticipated these ideas.

One modern attempt to realize something like Leibniz's dream is the software package *Mathematica* and its associated online compo-

---

[8] Leibniz, "De Progressione Dyadica, Pars I," MS, 15 March 1679

nent called *Wolfram Alpha*. Along with the ability to perform symbolic algebra, it also contains an enormous amount of curated statistical data in diverse fields. Stephen Wolfram, who guides development of the software, is consciously trying to realize his own interpretation of Leibniz's ideas, calling the *characteristica universalis* the "closest precursor" to the software.

5. A mechanical system to perform these logical operations.

Figure 41: The Logical Piano

Leibniz built the first mechanical calculator capable of multiplication and division. Up until that time *calculator* had been solely a job description. Scholars are just beginning to recognize the depth of Leibniz's understanding of the nature of computation[9] and its relationship with

---

[9] He also was one of the first to suggest looking at fractals as a way that nature could take a simple rule and build life forms. All life, he suggested, is built of small pieces which contain within them smaller elements of life, machines made up of smaller machines *ad infinitum*. His main reason for thinking this was observing that the body was made up of cells, and that sperm were also cells, and yet somehow contained within them the information needed to reconstruct a full body. This idea of recursion was later explored by Lady Ada Lovelace in one of the first computer programs. She believed that algebraically generated patterns might repay the favor to weaving which had contributed so much to the design of Babbage's mechanical computer.

the universe. He anticipated one of the central ideas in information theory:

> In Sections V and VI of his Discourse on Metaphysics, Leibniz asserts that God simultaneously maximizes the variety, diversity and richness of the world, and minimizes the conceptual complexity of the set of ideas that determine the world. And he points out that for any finite set of points there is always a mathematical equation that goes through them, in other words, a law that determines their positions. But if the points are chosen at random, that equation will be extremely complex.[10]

The ability to perform logical operations is at the heart of what we call a computer. But fundamentally, it is not a particularly difficult operation, nor are the outputs very profound. In the late 1700s, the Earl of Stanhope developed a device he called "The Demonstrator" for returning the results of a logical deduction. It was essentially a cleverly designed look-up table. This inspired Stanley Jevons to build the much more ambitious "Logical Piano" in 1869. Of its invention he wrote:

Figure 42: John Venn

> As I awoke in the morning the sun was shining brightly into my room, there was a consciousness on my mind that I was the discoverer of the true logic of the future. I felt a delight such as one can seldom hope to feel. I remembered only too soon though how unworthy and weak an instrument I was for accomplishing so great a work and how hardly I could expect to do it.[11]

---

[10] Gregory Chaitin, "Epistemology as Information Theory: From Leibniz to Ω."

[11] C Black and R Konekamp, *Papers and Correspondence of William Stanley Jevons*, Volume 1, "Journal of William Stanley Jevons" for 28 March 1866, MacMillan Press, 1973, p. 204

It was called a "piano" because it had an input keyboard, including a *finis* key that doubled as "Enter" and "reset." Given the input "A = AB and B = BC finis" typed into the keyboard, it would, through a system of levers and pulleys, return that the possible states include ABCD and ABC ~D, but not, for example, A~BCD. (The symbol ~ means "not.")

However, the inventor of the Venn diagram, John Venn, pointed out that the same thing could be worked out on pencil and paper without significantly more trouble. Jevons himself admitted that the piano wasn't of much use besides as a teaching device.[12] The real trick of logical thinking isn't in the performing of the deductions, but in the rest of Leibniz's project. Venn wrote:

> I have no high estimate myself of the interest or importance of what are sometimes called logical machines, and this on two grounds. In the first place, it is very seldom that intricate logical calculations are practically forced upon us; it is rather we who look about for complicated examples in order to illustrate our rules and methods. In this respect logical calculations stand in marked contrast with those of mathematics....
>
> In the second place, it does not seem to me that any contrivances at present known or likely to be discovered really deserve the name of logical machines. It is but a very small part of the entire process, which goes to form a piece of reasoning, which they are capable of performing. For, if we begin from the beginning, that process would involve four tolerably distinct steps.
>
> There is, first, the statement of our data in accurate logical language. This step deserves to be reckoned, since the variations of popular language are so multitudinous, and often so vague and ambiguous, that they may need careful consideration before they can be reduced to form.
>
> Then, secondly, we have to throw these statements into a form fit for the engine to work with—in this case the reduction of each proposition to its elementary denials. It would task the energies of a machine to deal at once, say,

---

[12] Jevons was also the inventor of the economic theory of utility. See chapter IX on economics for more about Jevons.

with any of the premises employed even in the few examples here offered.

Thirdly, there is the combination or further treatment of our premises after such reduction.

Finally, the results have to be interpreted or read off. This last generally gives rise to much opening for skill and sagacity... I cannot see that any machine can hope to help us except in the third of these steps; so that it seems very doubtful whether anything of this sort really deserves the name of a logical engine. [13]

The overall system formed of these five projects is a familiar one. When we program computers, English, a natural language, is translated into a precise computer language (such as C or Fortran) with simplified grammar and limited vocabulary. This in turn is transformed into binary notation (which Leibniz also invented for this purpose[14]) and fed into a device which performs logical operations on it. The result is that processes which were previously performed by human thought (arithmetic, proofs, if/then statements) can be performed automatically. Leibniz suggested that "all disputes could one day be settled with the words 'Gentlemen, let us compute!'"[15]

---

[13]John Venn, *Symbolic Logic,* 1881

[14] Leibniz saw a symbol of the *ex nihilo* creation of world in binary notation. He wrote:

"This is the origin of things from God and nothingness, positive and privative, perfection and imperfection, value and limits, active and passive, form and matter which is itself inactive...I have made those things clear to some extent by the origin of numbers from 0 and 1, which I have observed is the most beautiful symbol of the continuous creation of things from nothing, and of their dependence on God. For when the simplest progression is used, namely the dyadic [base 2] instead of the decadic [base 10] or the quaternary [base 4] all numbers can be expressed by 0 and 1." (letter to Johann Schulenburg 29 March 1698)

[15] This idea, that mathematics could someday subsume all truth, was believed by many mathematicians. Russell Whitehead (author of the *Principia Mathematica* which essentially brought all of mathematics then known into an axiomatic system) wrote:

"The ideal of mathematics should be to erect a calculus to facilitate reasoning in connection with every providence of thought, or external experience, in which the succession of thoughts, or of events can be definitely ascertained and precisely stated. So that

In the subsequent development of AI, a surprising reversal took place. Since Aristotle, man had been defined as "a rational animal." What made man special, the root of his intelligence, was his ability to use reason. A beast of burden could be mechanized and, in theory at least, be indistinguishable in behavior from the real thing. But man, with his ability to think rationally, could not.

With the developments of mechanized logic by Boole, Leibniz, and Babbage, the idea of a machine able to reason seemed possible. This attitude, that applying the rules of reasoning to terms was the key to thinking, continued through the development of AI systems, and was the reason for much early optimism in the field. And the mechanization of logical reasoning *was* successful. But it gradually became clear to nearly all researchers that even a sound system for reasoning and an enormous database of facts to reason on would not be enough to create an intelligent system.

Leibniz and these early AI researchers concentrated on the fact that logical reasoning from solid premises could lead to solid conclusions. One great difficulty in practically applying this is that the number of possible conclusions that can be reached is so astronomical. In the beginning, Llull saw this combinatorial explosion as a virtue, showing the ability of the system to grow beyond its inputs. Unfortunately, this enormous number of possibilities makes searching for a useful path from premises to conclusion too time consuming to be practical in real-life situations (as opposed to game worlds, like chess, where such approaches have proved successful as computers became sufficiently powerful.)

---

all serious thought which is not philosophy, or inductive reasoning, or imaginative literature, shall be mathematics developed by means of a calculus." (Whitehead, *Principia Mathematica*, 1898) This hope would be crushed by Kurt Godel's proof that no axiomatic system can ever hope to include all true theorems about that system. Today, mathematicians understand that because of this, mathematics will always be open to creative new ideas that open up new axiomatic systems.

Reasoning had been seen as the pinnacle of human mental achievement. But it turned out that it for machines, reasoning to valid conclusions was the easy part. Gathering information about the world, recognizing it, and deciding what actions to take based on that information (tasks so simple that insects can do them) proved to be much more difficult to automate.

Once the ability to perform reasoning had been automated, it became clear that what a mathematician does when proving a theorem is quite similar to what an artist does, a creative process. There are leaps of inspiration, recognitions of previously unseen metaphors, inspired guesses and mental images created and manipulated. Through these processes a path to the conclusion is discovered. The finished piece, the proof, gives very little clue about how future proofs might be created.

In this way, a type of creativity is part of all kinds of daily tasks. Rather than being opposed to reason, it is a fundamental and necessary part of how reason can be made to work.

VIII
The True But Unattempted Way:
Automated Induction

> There are and can exist but two ways of investigating and discovering truth. The one hurries on rapidly from the senses and particulars to the most general axioms, and from them, as principles and their supposed indisputable truth, derives and discovers the intermediate axioms. This is the way now in use. The other constructs its axioms from the senses and particulars, by ascending continually and gradually, till it finally arrives at the most general axioms, which is the true but unattempted way.
>
> *Novum Organum*, by Francis Bacon

The techniques of mechanical logic developed by George Boole were essentially deductive methods. When using deductive logic, we are given a full set of the rules to be followed from the outset, and follow these rules to their inevitable conclusions. Deduction is mathematically rigorous, and conserves what truth we have, guaranteeing that statements derived from true axioms are themselves true. Deductive reasoning is an exceptionally powerful tool, and was recognized as such since Aristotle spelled it out. As Bacon points out in the quote above, deductive logic was well understood and in wide use in his day.

However, deductive reasoning can only take us from the general to the particular. Knowing that "all men are mortal" and that "Socrates is a man" we can reason that "Socrates [in particular] is mortal." The question arises: how do we get true axioms in the first place? Presumably, we know that all men are mortal *because* every man yet observed has eventually died. This is reasoning in the opposite direction, though: from the individual facts to the general theory. Could such a reverse approach be made as rigorous and reliable as deductive reasoning?

108

Inductive Reasoning

The trouble with inductive reasoning, as this is called, is that a single counterexample serves to invalidate the truth of the conclusion. Since we have difficulty observing all men, we don't know whether one of the ones we missed might in fact be immortal. (There have been claims to that effect regarding a certain member of the MacLeod clan, for example.) The best we can say is that, at least as far as we know, most men are mortal. And we can conclude from this using deductive reasoning that Socrates might be mortal (though we aren't certain of that).[1] This kind of watered down conclusion was unpalatable, and numerically rigorous probabilistic reasoning wasn't really possible until the development of probability theory. The study of probability only began in the year 1654, when the mathematicians Pierre de Fermat and Blaise Pascal began to examine the concept of probability in gambling games of chance.

Figure 43: Francis Bacon

It also takes a lot more work to use induction. The scientist needs to go out into the world and collect a lot of examples before beginning to reason from it. It's not something one can do just sitting in an armchair. In fact, many of the great classics of deductive reasoning in Western writings today seem fundamentally flawed because they started from flawed premises about the nature of the world and of the mind. Francis Bacon wrote on this theme in *Novum Organum*, which proposes induction as a way that science can move forward:

---

[1] After writing this paragraph, I found that Bertrand Russell made this exact same argument with the same canonical example (though he didn't mention *Highlander*.) (Russell, *Problems in Philosophy*, 1912)

> [Philosophers] sought some support for the mind, and suspected its natural and spontaneous mode of action. But this is now employed too late as a remedy, when all is clearly lost, and after the mind, by the daily habit and intercourse of life, has come prepossessed with corrupted doctrines... The art of logic therefore being too late a precaution ...has tended more to confirm errors, than to disclose truth. Our only remaining hope and salvation is to begin the whole labor of the mind again; not leaving it to itself, but directing it perpetually from the very first, and attaining our end as it were by mechanical aid.

Bacon is proposing that by using newly invented scientific instruments to precisely measure phenomena, we could discover general principles in a rigorous way, instead of simply stating them relying on intuition, common sense, and scripture as was the practice among philosophers up to that point. The book greatly influenced the Royal Society in England, who helped develop the scientific method as we know it today—they saw themselves as carrying out Bacon's method.

Bacon was deeply influenced by the medieval ideas about the art of memory and the categorization of the world. He explicitly rejected the magical, cabbalistic, and showy aspects that had attached themselves to this tradition and focused on how finding regularities in nature was the key to understanding it. In particular, to understand the nature of human feelings and senses, one needed to study their artificial counterparts:

> The precepts followed by musicians, of passing from harsh and dissonant harmonies to sweet and consonant harmonies, is this not just as true for the emotions?...Do not the organs of the senses not perhaps have an affinity with instruments of reflection: the eye to the mirror, the ear to a narrow, concave instrument? These are not similitudes...but the signs and traces of nature which are imprinted in diverse matters and subjects.

By induction, Bacon believed, the axioms underlying the order of nature could be discovered. Induction seems to be the only way of introducing new knowledge into our system that wasn't there before, at least implicitly. It seems to more closely match the way that science works in practice, especially in fields such as biology or sociology.

Inductive and Deductive AI

Two approaches to AI have grown out of these two methods of formalized thinking. The method discussed in the previous chapter is known as "Top Down" AI. It includes such ideas as *expert systems*. In these systems, the rules are handed to the AI by the programmer explicitly and it combines them appropriately to reach conclusions.

The other strand of work includes neural nets and machine learning techniques. "Bottom Up" AI starts with a large body of data and either tries to formulate simpler rules to explain it, or, to explicitly estimate a conclusion without creating a simplified model at all.

We can compare the two approaches most easily when they are both used to solve the same problem. One example of this is natural language translation. The first approaches that computer scientists tried was to take written text in one language, and used hand-coded rules of grammar to try to diagram the sentences, breaking them up into subject, verb and object phrases according to firm rules with explicit exceptions where necessary. Then the words were translated and reassembled using the corresponding grammatical structure in the other language.

That was the idea, anyway—it didn't work out very well in practice. The researchers found that it took more and more rules and exceptions to those rules to get a few percent improvements in accuracy.

Contrast that with the approach Google used for translation. They collected an enormous body of bilingual texts, texts that had already been translated by hand into the target language. Then they simply looked for each phrase to be translated, and used the translated version of that phrase in the output. The results were surprising: even such a simple approach outperformed the best that hand-coded top-down approaches were able to do by a wide margin. The rules of the grammar were implicit within the patterns of how people actually used words.

This goes beyond mere induction. In inductive science, one tends to infer a relatively simple rule from a set of examples. Then that rule is applied to the new cases. But Google's method skips the rule, applying the examples directly to the specific case. This transduction allows for a rule far more complex than could be concisely stated.

In many different fields, the results have been the same: working bottom up from a lot of data beat out algorithms designed from the top down using relatively little data. One result of this is that we no longer understand exactly what it is that the machine is doing. In top-down systems, predicting the output is like predicting the path of a particle under the influence of various fields. The arrangement may be complex, but the interaction with each field is completely understood. Bottom-up systems can be more like negotiations where the decision must be made by consensus—the jury in *Twelve Angry Men*, for example. Ideas are proposed and considered; some may gain momentum, while others are overcome by an alliance of other factions. The talk goes back and forth until a decision is reached that is the most acceptable to as many people as possible. It is a chaotic and organic kind of interaction, and we have reason to believe it reflects what happens within our own brains.

Postmodern Computing

Another way to look at this is as a shift from an Enlightenment philosophy project to a postmodern one. Leibniz's project was very much a part of the Enlightenment, the Age of Reason: a massive encyclopedia, a reformed logical language that works by pure reason, the assumption that there is one truth that can be captured by science and express everything of interest, even the Latin style are very much part of the spirit of those times. This same philosophy was still held by the mathematicians, engineers, and scientists who built the early AI projects and expert systems.

A system that learns by example is, in contrast, postmodern. It doesn't assume there is one right way, but tries to capture variation and diversity. The results are a collage, a pastiche, of tiny quotations from other works. It grows organically and asymmetrically. It doesn't claim "correctness" or some minimal representation, but instead comes up with a pragmatic solution from a variety of elements that only make sense when considered as a whole. Instead of laying down the law about how things ought to be, it simply tries to capture what aspects of reality it has encountered. In these ways artificial intelligence research has followed the same shift in fashion that swept through architecture, writing, and the arts.

Both kinds of approaches can be described in mathematical language, and are at some level programmed in computer languages which are rigorously formal. So in a sense, the postmodern is only able to support itself in an environment of rationality. This can fairly be said of other art forms as well: Postmodern architecture, underneath the surface, uses rigid engineering principles to support the structures. Postmodern textual criticism and deconstructionism can only function in an environment which has first been constructed, using the tools of reason against itself.

Simulation

The new approach can also be characterized by a step-by-step simulation of a system as opposed to direct calculation of the end state of that system. For example, the amount of fluid flow in a pipe after running for five minutes can under some conditions (when the water is flowing slowly and smoothly) be calculated using the Navier-Stokes equations. But in order to handle the complex and chaotic fluid flow that occurs when the water is turbulent, the only option is to explicitly calculate the changing state of the water at successive time steps, knowing that the result will only be a typical motion of the water, not capturing the exact motion of each drop of water (whose motion is sensitive to initial conditions). Simulation, too, is an aspect of postmodernism. Jean Baudrillard, an important figure in the development of postmodernism, wrote *Simulation and Simulacra*, arguing that simulation can in some cases become indistinguishable from reality; that in fact the reality of social systems is built entirely from layer after layer of the artificial.

What we mean by creative depends a lot on the cultural conditions, on what has already been done, on understanding the nature of the human condition and the current conception of art. The only way this could be put into a machine is either explicitly—in which case it is the programmer rather than the machine who deserves credit for most of the creativity—or by learning from examples. A program can learn a lot about what we mean by beautiful by considering examples of things that have previously been determined to be beautiful.

Postmodern critics argue that the artist's intention in creating a work is not very useful in understanding the meaning of the work. This is rejecting one model of how creativity works (as a series of conscious rational steps and decisions) and promoting another (a flow of unconscious recombination of ideas within the mind of the artist, only partially understood and observed by the artist in much the same way it would be observed by any independent observer).

The idea that creativity consists in combining previously existing parts was explored by the philosopher John Locke in *An Essay Concerning Human Understanding*. What he calls *simple ideas* we would call qualia: the examples he gives include the color red and the taste of a pineapple, ideas which cannot be communicated by language except to those who have already experienced them. These simple ideas are combined together by the mind to make *complex ideas*:

> These essences of the species of mixed modes are not only made by the mind, but made very arbitrarily, made without patterns, or reference to any real existence.

The means by which this happens are not explained, except to say that it is guided by reason for the purposes of communication. Until we have some theory for how these combinations are formed, the theory of creativity as recombination of existing ideas is problematical. If a new idea consisted only of old ideas, we wouldn't call it creative. In a creative idea, what has been newly created is the information that these particular ideas, when combined together, have a value, or a beauty, or an affinity that was previously unrecognized. If that is true, perhaps what is most needed for machines to be creative is to somehow give them a sense of beauty or value that transcends what is already known to be good and extends to recognize whether new things are interesting as well. The inductive process could be an effective way to build such a model.

IX
Darwin Among the Machines:
Creative Evolution and the Invisible Hand

> As every individual, therefore, endeavors as much he can both to employ his capital in the support of domestic industry, and so to direct that industry that its produce may be of the greatest value; every individual necessarily labors to render the annual revenue of the society as great as he can. ... [H]e is in this, as in many other cases, led by an invisible hand to promote an end which was no part of his intention. Nor is it always the worse for the society that it was not part of it. By pursuing his own interest he frequently promotes that of the society more effectually than when he really intends to promote it.
>
> – Adam Smith, The Wealth of Nations

The Invisible Hand

Adam Smith realized that markets were capable of coming up with answers to questions that no human had the ability to solve. It was as if the system itself were exhibiting intelligent behavior. This is what he referred to in 1776 as the "invisible hand." Today, scientists would call this effect emergent behavior in a complex adaptive system.

Figure 44: Adam Smith

But this behavior posed a problem—how did it happen that the world would be arranged in such a way that individuals seeking their own ends should happen to lead to such a happy state? For Adam Smith, the answer was simple: God designed it that way. Leaving aside the theological question, however, what he proposed was perhaps the first realization of the idea of a creative rational mechanism: a designer could create a system that was capable

116

of coming up with rational and creative decisions through an arrangement of many components, each following simple rules.

> Smith's identification of the processes associated with the unintended consequences of individual actions in such diverse phenomena as language, money, moral sentiments, exchange and markets, across social experience, are usefully judged to be an early recognition of evolutionary "emergent order."[1]

Adam Smith was interested in the idea of the universe as a machine, and wrote about it when discussing the sect of the Pythagoreans:

> Mind, and understanding, and consequently Deity, being the most perfect, were necessarily, according to them, the last productions of Nature. For in all other things, what was most perfect, they observed, always came last. As in plants and animals, it is not the seed that is most perfect, but the complete animal....
>
> As soon as the Universe was regarded as a complete machine, as a coherent system, governed by general laws, and directed to general ends, viz. its own preservation and prosperity, and that of all the species that are in it; the resemblance which it evidently bore to those machines which are produced by human art, necessarily impressed those sages with a belief, that in the original formation of the world there must have been employed an art resembling the human art, but as much superior to it, as the world is superior to the machines which that art produces... According to Timaeus...that intelligent Being, who formed the world, endowed it with a principle of life and understanding, which extends from its centre to its remotest circumference, which is conscious of all its changes, and which governs and directs all its motions to the great end of its formation. This Soul of the world was itself a God, the greatest of all the inferior, and created deities...[2]

In other words, the world was seen as similar to a constructed machine, and that machine had been built to have consciousness and

[1] Gavin Kennedy, *Adam Smith and the Invisible Hand*
[2] Adam Smith, *Essays*, 1869, chapter IV (*History of Ancient Physics*) p. 392

understanding. This was an early expression of the idea that under-standing could be somehow constructed.

Stanley Jevons and Alfred Marshall

Figure 45: Stanley Jevons

Other early economists were also inter-ested in mechanical theories of the mind. Stanley Jevons has already been mentioned as the builder of the logical piano, but he is much better known for his pioneering work on economics and statistics, as the inventor of the theory of utility. At 19, during the gold rush in Australia, he was the assayer for the mint in Sydney, checking the quality of the gold being used to make coins. Like Isaac Newton, who was in charge of the mint in England, this put him in an unusual position as a scientist deeply involved in economic issues. He also was an avid photographer when the field was still quite new. He documented the life of the city as it was forming, and so was able to watch as an economic system formed around him. Jevons saw similarities between the interior workings of the human mind and the economic system.

His colleague, Alfred Marshall, also developed one of the first theories of how the mind is able to learn and adapt to its surround-ings. At the time, of course, few machines were able to adapt in any way, so understanding how the brain could change, learn and grow and yet still be explained mechanically was a real challenge. As part of his work to make economics a mathematical discipline, he needed a kind of model for an individual's reasoning that was simple enough to be predictable but complex enough that it could capture adaptive behavior. What he hit on was surprisingly similar to the theory behind neural networks, long predating Hebb's neuron.

Marshall wrote "While ... the Human organism may be likened to a keyed instrument, from which any music it is capable of produc-ing can be called forth at the will of the performer, we may compare a Bee or any other insect to a barrel-organ, which plays with the greatest exactness a certain number of tunes that are set upon it, but can do nothing else."

In *Ye Machine*, written during the 1860s, he asked his reader to imagine a machine that contains many spinning disks, each on its own axis and connected to other disks by elastic bands. The system would be used to generate responses to inputs. When the responses were appropriate, the machine would have a positive response, indicating that the bands between disks involved in the response would be tightened. If the responses were inappropriate, these bands would be loosened, making such a response less likely in the future. He

Figure 46: Alred Marshall

also used the metaphor of a channel cut by water; reinforcing the path in such a way that more water flows there in the future. He believed that this ability to adapt would lead to a machine that was less determined in its actions. It would be influenced by its character, which had been built up over time, but respond to each new situation in a flexible way. He understood that this would involve what we today call feedback loops, and that new levels of behavior would build on functions already developed. (He realized that such a machine would be limited in its problem solving ability only by speed and memory, and that a sufficiently large machine built on such a design could, for example, play a perfect game of chess.)

Order through self-organization was a major discovery of economists. Engels noticed how it had shaped the streets of Manchester—the city had grown quickly with the advent of industrialization, and its growth was completely unplanned, but self-interested principles had resulted in the city being divided sharply into rich and poor areas.[3]

All of these early economists saw that the same structures that caused markets to solve problems automatically might also be used by components within an individual mind to solve problems in much the same way.

---

[3] Steven Johnson, *Emergence*, p. 37

Gradual, steady, unerring, deep-sighted selection

Darwin's theories of evolution were similar to these economic theories in spirit. Like Smith, Jevons, and Marshall, Darwin was trying to explain how order appeared in the world spontaneously. What Darwin postulated was a system of comprehensible laws that brought about new, adapted species creatively. In *The Origin of Species* he spoke of this creative action metaphorically, in personified terms:

Figure 47: Charles Darwin

> [Nature] cares not for mere external appearance; she may be said to scrutinise with a severe eye, every nerve, vessel & muscle; every habit, instinct, shade of constitution,—the whole machinery of the organisation. There will be here no caprice, no favouring: the good will be preserved & the bad rigidly destroyed.... Can we wonder then, that nature's productions bear the stamp of a far higher perfection than man's product by artificial selection. With nature the most gradual, steady, unerring, deep-sighted selection,—perfect adaption [sic] to the conditions of existence.... [4]

The conclusion of *The Origin of Species* also deals with this theme. Darwin suggests that the Creator designed the world such that the birth and death of species, like the birth and death of individuals, is caused by commonplace accidents of history, rather than miraculous exceptions.

> To my mind it accords better with what we know of the laws impressed on matter by the Creator, that the production and extinction of the past and present inhabitants of the world should have been due to secondary causes, like those determining the birth and death of the individual...

> And as natural selection works solely by and for the good of each being, all corporeal and mental endowments will tend to progress towards perfection.

---

[4] Darwin 1856, in Stauffer *Charles Darwin's "Natural Selection,"* 1974, p. 224–25

It is interesting to contemplate a tangled bank, clothed with many plants of many kinds, with birds singing on the bushes, with various insects flitting about, and with worms crawling through the damp earth, and to reflect that these elaborately constructed forms, so different from each other, and dependent upon each other in so complex a manner, have all been produced by laws acting around us... Thus, from the war of nature, from famine and death, the most exalted object which we are capable of conceiving, namely, the production of the higher animals, directly follows. There is grandeur in this view of life, with its several powers, having been originally breathed by the Creator into a few forms or into one; and that, whilst this planet has gone cycling on according to the fixed law of gravity, from so simple a beginning endless forms most beautiful and most wonderful have been, and are being evolved.[5]

Like Adam Smith, Darwin had proposed a system of enumerated laws that once having been laid down are capable of generating entirely new forms on their own. Darwin's mechanism to explain how gradual change in life over time was able to come about had a profound effect on thinkers of the time. Evolution was of interest not only because it explained the diversity of life, but because it was one of the first plausible explanations of how creativity and innovation were possible.

Mechanism regards only the aspect of similarity or repetition. It is therefore dominated by this law, that in nature there is only like reproducing like. The more the geometry in mechanism is emphasized, the less can mechanism admit that anything is ever created, even pure form. In so far as we are geometricians, then, we reject the unforeseeable. We might accept it, assuredly, in so far as we are artists, for art lives on creation and implies a latent belief in the spontaneity of nature. But disinterested art is a luxury, like pure speculation. Long before being artists, we are artisans; and all fabrication, however rudimentary, lives on likeness and repetition, like the natural geometry which serves as its fulcrum. Fabrication works on models which it sets out to reproduce; and even when it invents, it proceeds, or imagines itself to proceed, by a new arrangement of elements

---

[5] Charles Darwin, *The Origin of Species*, 1859

already known. Its principle is that "we must have like to produce like."[6]

Darwin among the Machines

Living things seem to be very different than non-living things. Living things can move themselves. They can repair themselves when damaged. They can sense and react to the world. Their chemistry seems different than inorganic chemistry. Most miraculously of all, they can reproduce.

Figure 48: Samuel Butler

For centuries, many philosophers were preoccupied with the question of what it means to be alive. Many assumed that living matter was a fundamentally different kind of stuff than non-living matter. The idea of *vitalism* only slowly gave way to our mechanistic understanding of the fantastically complex processes within living cells. What seemed to be fundamentally different was merely much more complex. This gives cognitive scientists hope that the mysteries of the mind will someday be resolved in a similar way.

Although we now understand how life works in a certain sense, it far surpasses what we are able to engineer. No machine has yet been built that can repair or reproduce itself. Even digestion and photosynthesis, comparatively simple chemical reactions, have not been reproduced in a way usable by our machines. We are gradually making progress on each of these fronts. The RepRap project, for example, is an attempt to build a 3-D printer that can print out all its own parts, a form of self reproduction. The original promise of nanotechnology was the possibility of building a universal assembler, a tiny factory capable of building anything we could describe directly out of atoms, including a copy of itself. The necessary components are gradually being developed, but a working system still seems decades away.

---

[6] Henri Bergson, *Creative Evolution*, 1911

All of this complexity—for the moment, at least, superhuman complexity—demanded an explanation. Evolution gave a mechanistic explanation for how such a phenomenally sophisticated system could come about naturally.

Samuel Butler

Samuel Butler was the son of a pastor, and (like Sir David Brewster in the previous chapter) studied to become one himself at Cambridge. But he began to have doubts about his faith before his ordination, and perhaps to avoid the anger of his father, he emigrated to New Zealand. While there he was able to observe what was effectively the formation of a new society. It gave him a unique perspective on how the rules of society form, and how they might be different. He was especially interested in the relationship between society and new technologies.

Darwin's theory had an enormous impact as soon as it was published on scientists around the world. Even in New Zealand, which was a journey by ship of months from Europe, it was to have a deep effect on the way the natural world was understood.

The Origin of Species was published in 1856, and by the next year Samuel Butler published *Darwin among the Machines*, which he later elaborated into a section of the Utopian novel *Erewhon*. One of the first works of science fiction, it proposes the idea that evolution need not be restricted to the animal kingdom, but could be put to use in developing technology that would be able to emulate human abilities, and even surpass them. In the story, the humans eventually realize the growing threat the machines pose, and destroy them entirely, making do with only the simplest tools. Along the way, however, Butler explores a wide variety of themes related to artificial creativity for the first time. In the long selections that follow, Butler addresses the possibility of machine consciousness, senses, language, reproduction, and evolution.

> Consciousness, in anything like the present acceptation of the term, having been once a new thing... why may not there arise some new phase of mind which shall be as dif-

ferent from all present known phases, as the mind of animals is from that of vegetables?...

There is no security...against the ultimate development of mechanical consciousness, in the fact of machines possessing little consciousness now. A mollusc has not much consciousness. Reflect upon the extraordinary advance which machines have made during the last few hundred years, and note how slowly the animal and vegetable kingdoms are advancing. The more highly organized machines are creatures not so much of yesterday, as of the last five minutes, so to speak, in comparison with past time. Assume for the sake of argument that conscious beings have existed for some twenty million years: see what strides machines have made in the last thousand! May not the world last twenty million years longer? If so, what will they not in the end become?...

But who can say that the vapour engine has not a kind of consciousness? Where does consciousness begin, and where end? Who can draw the line? Who can draw any line? Is not everything interwoven with everything?

...There is a kind of plant that eats organic food with its flowers: when a fly settles upon the blossom, the petals close upon it and hold it fast till the plant has absorbed the insect into its system; but they will close on nothing but what is good to eat; of a drop of rain or a piece of stick they will take no notice. Curious! that so unconscious a thing should have such a keen eye to its own interest. If this is unconsciousness, where is the use of consciousness?

Shall we say that the plant does not know what it is doing merely because it has no eyes, or ears, or brains? If we say that it acts mechanically, and mechanically only, shall we not be forced to admit that sundry other and apparently very deliberate actions are also mechanical?...

...the answer would seem to lie in an inquiry whether every sensation is not chemical and mechanical in its operation? whether those things which we deem most purely spiritual are anything but disturbances of equilibrium in an infinite series of levers, beginning with those that are too small for microscopic detection, and going up to the human arm and the appliances which it makes use of? Whether there be not a molecular action of thought, whence a dynamical theory of the passions shall be deducible? Whether strictly speaking we should not ask what kind

of levers a man is made of rather than what is his tempera-
ment? How are they balanced? How much of such and
such will it take to weigh them down so as to make him do
so and so?...[7]

Butler's position today would be called *functionalism*. It is the argument that systems that behave in ways
similar to conscious minds are themselves
conscious. Notice the words "where is the
use of consciousness?" He is not particularly concerned with the problem of
subjective perception, as long as the
machine can act like a conscious being,
responding appropriately to stimuli. In
his day this was such a radical position
that, despite Butler's protest that he was
perfectly serious, the whole chapter was

Figure 49: William Paley

laughed off as a satire of Darwin, and is still treated as such by many
literary scholars today.

Intelligent Design

The "watchmaker argument" is an argument for what today is
called "intelligent design." The most famous statement of it[8] was by
the Christian apologist William Paley:

> In crossing a heath, suppose I pitched my foot against a
> stone, and were asked how the stone came to be there; I
> might possibly answer, that, for anything I knew to the
> contrary, it had lain there forever: nor would it perhaps be
> very easy to show the absurdity of this answer. But suppose
> I had found a watch upon the ground, and it should be in-
> quired how the watch happened to be in that place; I

---

[7] Samuel Butler, *Erewhon*, 1880

[8] The argument was not original to Paley. There are examples of
similar arguments involving timepieces from Cicero, Voltaire,
Descartes, and others. As the argument was copied over and over
again by different authors, small changes were introduced and
those that were successful in improving the argument were
copied by later authors. In this way, the argument can be said to
have evolved over time.

should hardly think of the answer I had before given, that for anything I knew, the watch might have always been there.... There must have existed, at some time, and at some place or other, an artificer or artificers, who formed [the watch] for the purpose which we find it actually to answer; who comprehended its construction, and designed its use.

The next section of *Darwin Among the Machines* takes this argument for intelligent design and turns it on its head. Butler contends that the watch itself has undergone evolution of a sort, involving artificial rather than natural selection. From his other works, it is clear that Butler is less interested in the mechanism causing evolution than the fact of information transfer from parent to child. Butler believed that the mechanisms of information storage in inheritance were the same mechanisms used in the mind. Although scientists in later years judged him to have been wrong on that point, recent discoveries have shown that methylization of DNA is a primary mechanism for storage of memories within the brain, so the two are more closely connected than was previously believed. DNA computing is also an active area of research. The passage is also interesting for noting the tendency towards miniaturization of computing components.

The present machines are to the future as the early Saurians to man. The largest of them will probably greatly diminish in size. Some of the lowest vertebrate attained a much greater bulk than has descended to their more highly organised living representatives, and in like manner a diminution in the size of machines has often attended their development and progress.

Take the watch, for example; examine its beautiful structure; observe the intelligent play of the minute members which compose it: yet this little creature is but a development of the cumbrous clocks that preceded it; it is no deterioration from them. A day may come when clocks, which certainly at the present time are not diminishing in bulk, will be superseded owing to the universal use of watches, in which case they will become as extinct as ichthyosauri, while the watch, whose tendency has for some years been to decrease in size rather than the contrary, will remain the only existing type of an extinct race.

... I fear none of the existing machines; what I fear is the extraordinary rapidity with which they are becoming something very different to what they are at present...

Alfred Marshall also expressed the evolutionary potential of his psychological machine (discussed at the beginning of this chapter) in terms of Paley's watch. He suggested that if such a machine were to make imperfect copies of itself, it would also improve over time through the action of natural selection. This evolution, he reasoned, might explain how instinct (a mental operation) could be evolved and passed down.

Those ears will no longer be needed

Butler uses the whistle of a train as an example of machine communication. He sees the system of *train operator + train* as a unit, where the functions of the train operator, already constrained to act in set ways to certain signals, will gradually be replaced by machines of increasingly "delicate construction."

As yet the machines receive their impressions through the agency of man's senses: one travelling machine calls to another in a shrill accent of alarm and the other instantly retires; but it is through the ears of the driver that the voice of the one has acted upon the other... There was a time when it must have seemed highly improbable that machines should learn to make their wants known by sound, even through the ears of man; may we not conceive, then, that a day will come when those ears will be no longer needed, and the hearing will be done by the delicacy of the machine's own construction?—when its language shall have been developed from the cry of animals to a speech as intricate as our own?

... Take man's vaunted power of calculation. Have we not engines which can do all manner of sums more quickly and correctly than we can?... In fact, wherever precision is required man flies to the machine at once, as far preferable to himself. Our sum-engines never drop a figure, nor our looms a stitch... This is the green tree; what then shall be done in the dry?...

The Body Politic

The analogy of a group of people with a living body appears in Plato's *Republic*. [9] Plato suggests that the components of an individual human mind are difficult to make out, but that in form they are the same as those governing a society of individuals. If we can understand how a society works, then we will understand how the parts of a mind work. He suggests that the main three components of the mind are the will, the rational part and the appetitve part (the same division which Freud later called the ego, the superego, and the id). The body that is guided by the state is what we now refer to as the "body politic." The theme was taken up by Paul in the New Testament and in Thomas Hobbes' *Leviathan*. Here it is used as an argument that a system of disconnected parts can come together to act in many ways like a living creature.

> It is said by some that our blood is composed of infinite living agents which go up and down the highways and byways of our bodies as people in the streets of a city. When we look down from a high place upon crowded thoroughfares, is it possible not to think of corpuscles of blood travelling through veins and nourishing the heart of the town? No mention shall be made of sewers, nor of the hidden nerves which serve to communicate sensations from one part of the town's body to another; nor of the yawning jaws of the railway stations, whereby the circulation is carried directly into the heart,—which receive the venous lines, and disgorge the arterial, with an eternal pulse of people. And the sleep of the town, how life-like! with its change in the circulation...

> ...Are we not ourselves creating our successors in the supremacy of the earth? Daily adding to the beauty and delicacy of their organisation, daily giving them greater skill and supplying more and more of that self-regulating self-acting power which will be better than any intellect?...

---

[9] Plato, *Republic*, Book II, 369

Let him think of a hundred thousand years

Finally Butler comes to artificial intelligence. He argues that free will is an illusion; that our own actions are determined based on our previous experiences and current stimulus.

> But I have heard it said, 'granted that this is so, and that the vapour-engine has a strength of its own, surely no one will say that it has a will of its own?' Alas! if we look more closely, we shall find that this does not make against the supposition that the vapour-engine is one of the germs of a new phase of life. What is there in this whole world, or in the worlds beyond it, which has a will of its own? The Unknown and Unknowable only!
>
> ... Let any one examine the wonderful self-regulating and self-adjusting contrivances which are now incorporated with the vapour-engine, let him watch the way in which it supplies itself with oil; in which it indicates its wants to those who tend it; in which, by the governor, it regulates its application of its own strength; let him look at that storehouse of inertia and momentum the fly-wheel, or at the buffers on a railway carriage; let him see how those improvements are being selected for perpetuity which contain provision against the emergencies that may arise to harass the machines, and then let him think of a hundred thousand years, and the accumulated progress which they will bring.

This last idea, the governor that regulates inertia, was the original example of *cybernetics*, the study of self-regulating systems which would eventually become synonomous with artificial intelligence. (From cybernetics we derive terms such as *cyberspace*.) Darwin's collaborator Alfred Wallace wrote in 1858:

> We have also here an acting cause to account for that balance so often observed in nature,—a deficiency in one set of organs always being compensated by an increased development of some others—powerful wings accompanying weak feet, or great velocity making up for the absence of defensive weapons; for it has been shown that all varieties in which an unbalanced deficiency occurred could not long continue their existence. The action of this principle is exactly like that of the centrifugal governor of the steam engine, which checks and corrects any irregularities almost before they become evident; and in like manner no unba-

lanced deficiency in the animal kingdom can ever reach any conspicuous magnitude, because it would make itself felt at the very first step, by rendering existence difficult and extinction almost sure soon to follow.[10]

A case can be made that it was this advancement in self-observing, self-modifying machines that inspired the concept of natural evolution, rather than the other way around. Perhaps the most likely explanation is that the ideas in biology and mechanics formed a beneficial feedback loop.

George Eliot

George Eliot was inspired by *Erewhon* and another utopian novel of the day, *The Coming Race*, to write a short exploration of her own ideas on the subject. While Butler believed in the development of machine consciousness, she was more concerned that the machines might take over, and yet be utterly *un*conscious:

> When I am told of... of a machine for drawing the right conclusion, which will doubtless by-and-by be improved into an automaton for finding true premises; I get a little out of it, like an unfortunate savage too suddenly brought face to face with civilisation, and I exclaim—

> "Am I already in the shadow of the Coming Race? and will the creatures who are to transcend and finally supersede us be steely organisms, giving out the effluvia of the laboratory, and performing with infallible exactness more than everything that we have performed with a slovenly approximativeness and self-defeating inaccuracy?'

Figure 50: George Eliot

> ...[T]his planet may be filled with beings who will be blind and deaf as the inmost rock, yet will execute changes as delicate and complicated as those of human language and all the intricate web of what we call its effects, without sensitive impression, without sensitive impulse: there may be, let us say, mute orations, mute rhapsodies, mute discussions, and no consciousness there even to enjoy the silence.

---

[10] Alfred Wallace, "On the Tendency of Varieties to Depart Indefinitely From the Original Type," 1858

She also explored the possibilities of machine evolution through natural selection. She jokes about electrical machines getting into disputes with those descended from the more respectable family of steel-cutters, but writes more seriously:

> If...machines as they are more and more perfected will require less and less of tendance, how do I know that they may not be ultimately made to carry, or may not in themselves evolve, conditions of self-supply, self-repair, and reproduction...?

> [a machine might]... by a further evolution of internal molecular movements reproduce itself by some process of fission or budding. This last stage having been reached, either by man's contrivance or as an unforeseen result, one sees that the process of natural selection must drive men altogether out of the field; for they will long before have begun to sink into the miserable condition of those unhappy characters in fable who, having demons or djinns at their beck, and being obliged to supply them with work, found too much of everything done in too short a time.[11]

The Adjacent Possible

This chapter has talked about economics, evolution, and creative art. The scientist Stuart Kauffman has made explicit something these 19th century authors were illustrating: what is possible in each of these fields cannot be specified beforehand.

By way of contrast, consider a classical physical system, like ten classical hydrogen atoms in a box interacting by means of the weak nuclear force. In this case, we know that the position of each of the atoms will be a three dimensional location, and the velocities can also be written with three values per particle. So everything about all the positions and velocities can be recorded with sixty numbers. Their interaction may be complex and chaotic, but these sixty numbers will always suffice to describe the situation. One can ask questions such as, "What arrangement of the atoms maximizes the total distance between all the pairs?" and, in principle at least, calculate an answer.

---

[11] George Eliot, *Shadows of the Coming Race*, 1887

Now suppose that one wished to perform a similar calculation for the economy. Pretend that we could figure out how good every person in the economy would be at any currently existing job. Then one could ask the question, "If everyone spends 40 hours a week at work, what assignment of people to jobs would result in the goods with the most value produced?" The trouble is that there are certain jobs, such as an entrepreneur or a developer, whose job it is to create new jobs or new goods. These creative professions actually expand the number of possibilities in the problem. In this case, the number of dimensions of the problem depends on the assignment of people to jobs in the first place, and that number grows over time.

In evolutionary biology, a problem one would want to know the answer to might be "What body shape would be optimal for a leaf-eating creature?" In order to answer this problem, one could attempt to define what possible body shapes would be considered: for example, by starting with the body of an ant, and setting the length of each body segment as a parameter. Actually, artificial life researchers have tried similar experiments, and have "evolved" designs better at achieving their goals. But in each such experiment, the results have been similar: after an initial period of adaptation, the evolution plateaus, and no further improvements are seen. This is very different from what we see in the real world. Evolutionary mutations don't just alter predefined parameters. They actually create new possible ways of succeeding that didn't exist before. When we define the rules of the world in which they are simulated, we make vast simplifications for computational efficiency. Somehow these simplifications have prevented the discovery of new solutions once the limited space has been explored.

Art shares this open, growing possibility space. New works of art, to be understood and valued as art, must build on an existing tradition or react to it. There is no limit to the ways this building can happen, except what someone finds interesting or beautiful. A kaleidoscope can produce infinite new designs, but because the space of possible designs is limited beforehand (the position and orientation of each of the colored chips of glass) the toy eventually becomes less interesting as we recognize the theme and the ways it can vary.

Field	Growing configuration space
Economics	Goods, jobs, services
Evolutionary biology	Life forms
Art	Artistic creations

Part of the difficulty is selecting an appropriate evaluation function. For life, this is survival until reproduction can occur. For art, it is the criteria of interest or beauty. How would we implement this criterion into a machine? We could program it or train it on what we had already found beautiful, but this would only allow it to recombine things that had been found beautiful in the past. We would need to give it the ability to see for itself that a new idea is interesting. There has, to date, been remarkably little work on how to achieve this.

Many creative evolution programs use a human in the loop to perform the evaluation of possibilities, choosing variations that are preferred. In this case, the computer could be seen a tool being used by the artist doing the choosing, rather than the machine itself acting as an artist.

Where does all this leave us? I think that these early thinkers about evolution, economies, and the mind touched on something very important to the question of how to automate creativity. They conceived of a way that a systematic process could create innovative solutions to problems, and could grow into something more complex and capable than it had been before. The process of adaptation through competition and selection that Darwin discovered allows us to go beyond the model of creativity based on random rearrangement by adding in a selective step that modifies the generative process. Such an evolutionary system, however, still needs an evaluation function that can tend to move the generated artwork in a direction we humans find pleasing, itself a very difficult problem.

X
The Engine Might Compose:
Charles Babbage's Analytical Engine

Figure 51: Charles Babbage

Charles Babbage was a hacker, in the original positive sense of the word. As a child, he would disassemble his toys to learn how they worked, and never really outgrew his love of complicated toys. In school, he engaged in the kind of elaborate practical jokes that are common today at MIT, and was overly fond of bad puns and wordplay. He loved defeating security systems and working with codes, cryptography, and lockpicks (this and his interests in math and language he shared with both Leibniz a century before and Turing a century later). He kept a collection of automata and showed them off to his friends. As a young man he developed an ideal, precise language:

> I accidentally heard, for the first time, of an idea of forming a universal language. I was much fascinated by it, and, soon after, proceeded to write a kind of grammar, and then to devise a dictionary. Some trace of the former, I think, I still possess: but I was stopped in my idea of making a universal dictionary by the apparent impossibility of arranging signs in any consecutive order, so as to find, as in a dictionary, the meaning of each when wanted. It was only after I had been some time at Cambridge that I became acquainted with the work of Bishop Wilkins on Universal Language.[1]

Babbage understood very well the mathematical and practical uses that his machine could be put to. He also realized the philosophical

---

[1] Charles Babbage, *Passages from the Life of a Philosopher*, 1864

implications of building a machine to do what had previously been solely mental functions, writing "The mechanical means I employed to make these carriages bears some slight analogy to the operation of the faculty of memory,"[2] and "The great object of all my inquiries has ever been to endeavour to ascertain those laws of thought by which man makes discoveries."[3] He even envisioned it being used for entertainment purposes:

> I selected for my test the contrivance of a machine that should be able to play a game of purely intellectual skill successfully ; such as tit-tat-to, drafts, chess, &c. I endeavoured to ascertain the opinions of persons in every class of life and of all ages, whether they thought it required human reason to play games of skill. The almost constant answer was in the affirmative. Some supported this view of the case by observing, that if it were otherwise, then an automaton could play such games. A few of those who had considerable acquaintance with mathematical science allowed the possibility of machinery being capable of such work; but they most stoutly denied the possibility of contriving such machinery on account of the myriads of combinations which even the simplest games included.

He conceived of a way to give the illusion that his machine was making a spontaneous change in operation, by having it keep adding one each operation until it got to a million, and then beginning to count by tens. While such a function was not particularly complicated, it demonstrated that induction might lead us to one set of beliefs about natural laws, but that such a belief might at any moment prove to have been unjustified. What appears to be a miraculous change in rules can in fact be seen as simply the mechanical execution of a more complex rule. Such functions, which follow one rule until a certain condition is reached and then switch to a new rule, are among the most interesting small Turing machines for mathematicians. In many cases, the small Turing machines with the richest behavior are those whose condition for moving into an end state can be shown to correspond to a difficult-to-prove conjecture in number theory.

---

[2] *Ibid.*, p. 63

[3] *Ibid.*, p. 452

This idea—that repeatedly applying the same rule to the output of the previous computation can give rise to unexpected behavior, helping us to understand something about the workings of the natural world that had been unexplored by earlier scientists—is the key insight of the theory of cellular automata, and lies behind the richness of fractals like the Mandelbrot set. It was new and startling to Babbage. He saw it as a way to possibly explain the discontinuities, what we would now call the "punctuated equilibrium," in the fossil record. Recursive functions like this are also how pseudo-random numbers are generated, and so lie behind the creativity of any software that uses "random" numbers.

It was difficult at the time to convey the potential of a universal computer. It is reasonable to suppose that if those in the government had really understood what such a machine could accomplish, the machines would have been completed, one way or another. Babbage tried to communicate this by different analogies and applications. For example, a visitor pointed out that up until that point, thinkers had only been concerned with the legislative branch of mathematics, and that what Babbage was proposing was to build the executive branch. The analogies of a person playing a game, or of the working out of natural law, or the functioning of a government, were ways to try to communicate to non-scientists just what it was he was trying to build. But these metaphors were also how Babbage perceived the possibilities of what mechanical computers might accomplish in the future.

Babbage was not an artist, and had no interest in generating creative artwork with his engines; in fact, he despised the music of organ grinders and other street musicians so much that he tried to get laws passed banning it. (The ban failed, and the street musicians made a special point of playing right outside his window for years afterwards, even when he lay on his deathbed.) Recognizing the artistic potential in his machines fell to Lady Ada Lovelace.

Figure 52: Ada Lovelace Byron

Ada Lovelace Byron

Ada was the daughter of Lord Byron, the poet. Her mother Annabella divorced Byron when she obtained proof that he had slept with his half-sister. Annabella accused Byron of being insane because of his celebrity lifestyle and determined that Ada would have the most dry, mathematical and logical education possible in order to avoid her growing up to be like her father. (This amused him, and he affectionately referred to Ada as "the princess of parallelograms.")

Ada, however, with the help of good teachers who encouraged her imagination, developed an appreciation for the aesthetic possibilities of technology. At the age of thirteen, for example, she designed a flying machine. She eventually became a teacher and respected as a mathematician. This gave her the opportunity to meet Charles Dickens, Charles Darwin, Michael Faraday, Augustus DeMorgan, and many other brilliant thinkers and scientists of her time.

To spread understanding of the potential of computers more widely, she translated a paper by L. F. Menabrea. Though it would have been difficult, as a woman, to get her own work published, she was able to sneak in her own ideas as commentary on the translation.

Ada had an erratic character, and was prone to opium-induced delusions of her own self-importance. Some historians feel her contributions to computer science have been overrated. She was, like her father, a celebrity after all, and the phenomenon of a celebrity doing mathematics at all was just as remarkable in the

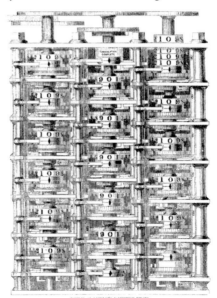

Figure 53: The difference engine

1800's as it is today (witness articles about the mathematical prowess of actress Danica McKellar or guitarist Brian May—what particular discoveries they have made is rather beside the point). Most are agreed, however, that her real contribution was largely in anticipating the philosophical implications of computational devices. In this respect her writings are similar to those of Mary Boole, George Eliot, or Mary Shelley. All of these women saw farther than their colleagues what the ramifications of these scientific advancements might mean for society and our understanding of life and the mind. The analogy that Babbage saw was mainly between the workings of his machine and natural law. Of her contemporaries, only Ada saw its potential as an artificial mind.

About Babbage's engines she wrote, "the Engine might compose elaborate and scientific pieces of music of any degree of complexity or extent...We may say most aptly, that the Analytical Engine weaves algebraical patterns just as the Jacquard loom weaves flowers and leaves."[4] In fact, after a visit to see such a loom, she persuaded Babbage to use a similar system of punched cards to program the engine he was designing.

But it is interesting to note that although she saw the potential for algorithmic composition, at the same time in her translation of Menabrea she denied the possibility of true understanding or creativity to the machine:

> T]he interpretation of formulæ and of results is beyond its province, unless indeed this very interpretation be itself susceptible of expression by means of the symbols which the machine employs. Thus, although it is not itself the being that reflects, it may yet be considered as the being which executes the conceptions of intelligence...

> The Analytical Engine has no pretensions whatever to originate anything. It can do whatever we know how to order it to perform. It can follow analysis; but it has no power of anticipating any analytical relations or truths. Its province is to assist us in making available what we are already acquainted with.

---

[4] L. F. Menabrea, translated by Ada Lovelace

For, in so distributing and combining the truths and the formulæ of analysis, that they may become most easily and rapidly amenable to the mechanical combinations of the engine, the relations and the nature of many subjects in that science are necessarily thrown into new lights, and more profoundly investigated. This is a decidedly indirect, and a somewhat speculative, consequence of such an invention.[5]

It seems that she saw in the engine a hope of understanding the bewildering complexity of her own mind. She wrote in a letter:

I have my hopes, and very distinct ones too, of one day getting cerebral phenomena such that I can put them into mathematical equations—in short, a law or laws for the mutual actions of the molecules of brain. I hope to bequeath to the generations a calculus of the nervous system.

Her goal was to understand the mind as a mechanism, but she was always keenly aware of the creative potential as part of the mind that might be automated. She saw the imagination as a critical part of intelligence, and tried to pass that on to her students.

Mathematical Science shows what is. It is the language of unseen relations between things. But to use & apply that language we must be able to fully appreciate, to feel, to seize, the unseen, the unconscious. Imagination too shows what *is*, the *is* that is beyond the senses. Hence she is or should be especially cultivated by the truly Scientific, — those who wish to enter into the world around us!

Babbage's engines were never successfully constructed; he ran out of money and time, not least because better ideas kept occurring to him for a new machine before the previous had been built. Through the rest of the 1800s, the story of Babbage's engines would serve as both a warning against trying to build ambitious computing devices and an inspiration that such a thing could be done in principle.

He realized that he was well ahead of his time. Long after his universal mechanical computer had failed to be completed, he wrote:

The great principles on which the Analytical Engine rests have been examined, admitted, recorded, and demonstrated. The mechanism itself has now been reduced to

[5] *Ibid.*

unexpected simplicity. Half a century may probably elapse before any one without those aids which I leave behind me, will attempt so unpromising a task. If, unwarned by my example, any man shall undertake and shall succeed in really constructing an engine embodying in itself the whole of the executive department of mathematical analysis upon different principles or by simpler mechanical means, I have no fear of leaving my reputation in his charge, for he alone will be fully able to appreciate the nature of my efforts and the value of their results.[6]

Machines Intellectuelles

The idea of punched cards for storing data was independently taken up by a Russian inventor, Semyon Korsakov, in 1832. Korsakov worked in the police ministry and kept extensive records on the populace. He devised a way of recording data by means of holes punched in cards. If two cards shared many of the same holes, it meant that the

Figure 54: Semyon Korsakov

data they contained was similar. This fact could be used for creating a kind of search engine, where a plate with pins would rapidly scan across hundreds of records, only falling in and stopping when the pins could fall through all of the holes, indicating an exact match. Korsakov was excited about his ideas, and thought that they could be used to enhance human intelligence in the same way that the microscope and telescope had been used to enhance human sight. He wrote,

> *machines intellectuelles* would limitlessly strengthen the power of our thought, as soon as distinguished scientists apply their knowledge to studying the principles of this process and compose the tables necessary for its application in various fields of human knowledge.[7]

---

[6] *Ibia.*, p. 450

[7] Monthly Lexicon (subject calendar) and General Staff of the Russian Empire for 1832. Part I. St. Petersburg

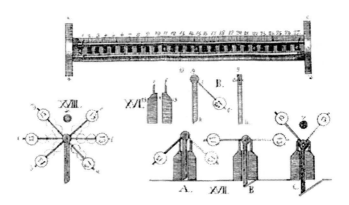

Figure 55: Design for ideoscope

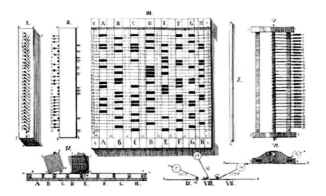

Figure 56: Design for linear homeoscope

His designs for the ideoscope and homeoscope were released to the world (open source fashion) rather than patenting them to encourage their widespread use and further development. Unfortunately the Russian Academy of Science didn't see the potential and little resulted from the inventions. He is today better remembered by the homeopathic medicine community for his remedies, than by the information science community for his ingenious method of searching through the database of those remedies. The representation of data in binary form on punched cards would later be used by Hollerith for the U.S. Census, popularizing their use for data storage by IBM in the early 1900s and leading to their use in electronic computers. It was only at this point that development of the field would seriously begin again, in the form of business machines, which were not capable of universal computation.

Once universal computers were finally built in the 1940s and 50s, the whole field of artificial creativity began to be much more widely explored. We are now making rapid progress, but as we climb each mountain we see that the peak we are trying to reach is even higher and more distant than we thought.

XI
To Fix These Fleeting Images:
Photography and Generative Art

The origins of photography begin with the observation that light passing through a small hole casts a projection. Aristotle, for example, noticed how during an eclipse the crescent shape of the partly covered sun was dappled on the ground, a projection through each of many tiny holes in the foliage overhead.

The camera obscura was first built by the Persian scientist Ibn al-Haitham around 1000 AD. At first these were simply darkened rooms with a pinhole for light to enter and cast an image onto the opposite wall, and eventually became a popular tool for artists. Wilhelm Homberg described the photochemical effect where light

Figure 57: The dappling of sunlight is circular because the sun is a circular light source. During an eclipse, the shape of these dapples changes. This illustrates the principle behind a pinhole camera. (Photo by Simon Cohen.)

changes the chemical structure of a compound in 1694, which is the chemical basis of photography.

Thus all the scientific pieces were in place for photography to be invented by 1700. It was not until the *idea* of photographs appeared in an early work of science fiction, however, that inventors began actively trying to create a photographic process. The novel *Giphantie*, which was published in 1760, describes "fixing" a photograph long before the first photographs were actually created in the 1820s:

> You know that rays of light reflected from different bo-
> dies form pictures, paint the image reflected on all polished
> surfaces, for example, on the retina of the eye, on water,
> and on glass. The spirits have sought to fix these fleeting

images; they have made a subtle matter by means of which a picture is formed in the twinkling of an eye. They coat a piece of canvas with this matter, and place it in front of the object to be taken. The first effect of this cloth is similar to that of a mirror, but by means of its viscous nature the prepared canvas, as is not the case with the mirror, retains a facsimile of the image. The mirror represents images faithfully, but retains none; our canvas reflects them no less faithfully, but retains them all. This impression of the image is instantaneous. The canvas is then removed and deposited in a dark place. An hour later the impression is dry, and you have a picture the more precious in that no art can imitate its truthfulness.[1]

Once the idea of a photograph had become widespread, there was a simultaneous effort to develop the technology by several independent inventors, including Niépce, Daguerre, and Talbot, but also others around the world.

Does photography count as machine-generated art? Some early authors called it "photogenic drawing." Clearly a machine, an invention, is somehow involved in the creation of a photograph. The second half of the question, "is photography art?" was a major topic of conversation and declarations for one side of the question or the other for the first few decades after its invention. For those who accepted that photography could be art, a big part of the discussion centered around the question of who, exactly, was the artist. It didn't seem like it could be the photographer, who was doing so much less work than traditional artists to get superior technical results. So was it the camera? The sun, personified as Apollo? Light itself? Nature? All of these were proposed by various writers, though more as a kind of poetic tribute than a serious speculation about non-human creativity.

Today, photographs have a place in art museums and art history textbooks, so the answer seems to be decided that at least some photographs can be art. And the place of the artist has certainly settled on the photographer. The photographer makes choices in how to set up the camera and the lighting, in setting the focus, filters, and lens distortion, and often arranges the elements to be photographed. Most

---

[1]Tiphaigne de la Roche, *Giphantie*, 1760

importantly, perhaps, the photographer chooses which photographs to present to the public. In all of these choices, the photographer has a chance to exercise creativity to attain an aesthetic effect, and that, we have come to believe as a society, is the role of an artist. The camera, for all its ability, is seen as little more than a paintbrush, a tool allowing the artist to express a vision.

Art and Creation

It wasn't until the development of mechanical means of reproduction of art—the printing press, photography, lithography, and the mechanical loom—that there was any need to make a distinction between the artist who created something original, and the artisan who created copies of the same work of art again and again. In ancient Greece, the word for art was *techne*, which means following rules, and from which we get words like *technology*. An artist wasn't supposed to create something new, but to be a master at following the strict art forms that had already been developed. In that sense, devices for the reproduction of art were considered to be automating one of the important roles of an artist. As has happened again and again in the history of automation, the essential function of the artist was then redefined to be something that machines could not yet do.

The word *creatio* was reserved for God alone until the 1800s. It was only then that artists began to refer to what they were doing as "creative," bringing into existence something original, and the idea of art was associated with originality rather than just skill. This idea met with some resistance at the time. Philosopher Denis Diderot, for example, believed that the imagination was composed of "the memory of forms and contents" and that it was only by rearranging, expanding on, or otherwise making recombination of experienced ideas that an artist made a work of art, but that nothing essentially new was being "created."

It would take another hundred years, until the turn of the 20th century, before the word "creative" would be applied to scientists and inventors. At first, this was done as an analogy with artists, until eventually the sense of the word broadened to include other groups of thinkers. Philosopher Henri Bergson's book *Creative Evolution* was perhaps the first work to take seriously the idea that non-human systems could be creative in this sense. In this book Bergson proposed that there was some kind of fundamental creative source (*élan vital*), closely related to consciousness, that drove life to evolve creative solutions to problems. While few scientists would agree with Bergson's philosophy today, the identification of human creativity and evolutionary creativity is becoming more popular as the use of genetic algorithms becomes more sophisticated.

Although this history of the usage of the word "creative" is true as far as it goes, it doesn't mean that people weren't thinking about creativity, just that they were using different words to describe it. During the medieval period, for example, the idea of human creativity was closely associated with memory, and the art of building a conceptual structure in the mind that would help in the creation of new ideas. The word *machina* was often used to refer (in an analogical way) to this mental construction that could generate new thoughts by recombining the ideas that were sitting in memory.[2]

Generative Art and the Accidents of Nature

The examples of artificial creativity in this book can be considered as early examples of "generative art." According to artist and scholar Philip Galanter, generative art is "any art practice where the artist creates a process, such as a set of natural language rules, a computer program, a machine, or other procedural invention, which is then set into motion with some degree of autonomy contributing to or resulting in a completed work of art." Photography doesn't have this autonomy—indeed, a lot of work has gone into eliminating every trace of it. Accidental effects in photography are usually seen as errors to be corrected.

---

[2] Mary Carruthers, *The Craft of Thought*, Cambridge University Press, 2000

Figure 58: Holga photograph by Matt Callow

Some photographic artists have experimented with allowing these effects a greater part in the finished image, deliberately choosing "imperfect" cameras, such as the cheaply produced Holga camera, to achieve a more natural effect. This is similar to Japanese Raku potters, who highly valued the accidental, fractal forms that occurred naturally when their pieces were fired, and resembled images from nature. The point for both of these kinds of artists is to let go of some of the control and allow beauty to happen on its own.

All of the creative machines discussed in this text use both whole elements, and ways of recombining them, in ordered or random ways (see the table at the beginning of the next chapter for a summary). Photography leans most heavily on the preexisting images in the world. In contrast, generative art focuses mainly on the generation process.

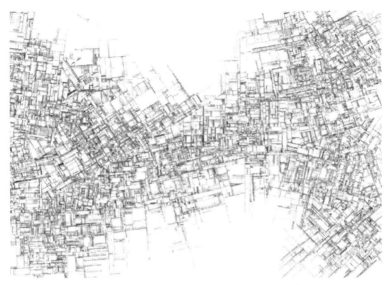

Figure 59: Image from the program *Substrate*, by Jared Tarbell

The term generative art is a fairly new one, encompassing stochastic art and algorithmic art, as well as others. In generative art, the artist sets up a process which is at least partially autonomous. The artist abandons some control to allow an algorithm or stochastic process to generate part of the artwork. Evolutionary algorithms, which try to replicate the process of evolution by using some aesthetic quality to stand in for fitness, have produced some of the most interesting results in this field.

Generative art is a synonym with artificial creativity. Perhaps the main difference is that the former is primarily pursued by artists interested in technology, while the latter is practiced by technologists interested in art. Innovators in both groups may find something of interest in this book to ground their work in the tradition of what has gone before.

Galanter has written extensively about generative art and its implications for creativity:

> All human forms of creativity, including creativity in the arts, are much more similar than not…Complex adaptive systems are those complex systems that both sense the changing nature of their surroundings and take actions to

maintain their existential integrity. These adaptive actions are examples of creativity. In lower life forms the adaptation, intelligence, and creativity involved may be quite basic. But they exhibit adaptation, intelligence, and creativity nevertheless. At the human level maintaining one's existential integrity involves congruity with social expectations, and creativity is always judged relative to a social context. Here social context can also be thought of as culture. Creativity in the arts is especially tied to culture, as its practical function is minimal...To the extent a computer can be considered a complex adaptive system it can also be considered creative.[3]

Figure 60: pens tied to the branches of a weeping willow by Tim Knowles. As the leaves blow in the wind, the pens create a pattern.

Another form of art created in this way is marbled paper, part of the art of bookbinding. Paper marbling entered Europe by way of Turkey around 1600, and was popular until the mid 19th century, when mechanical binding of books became practical. It appears to have first originated in China in the 10th century, as a method of preserving[4] the paper from worms and insects by dipping it in a chem-

---

[3] Philip Galanter, "Thoughts on Computational Creativity."

[4] On the connection between art and preservation, essayist Paul Graham writes, "jam, bacon, pickles, and cheese, which are among the most pleasing of foods, were all originally intended as methods of preservation. And so were books and paintings."

ical solution that happened to dye the paper as well. In Japan, this had developed into the art of *suminagashi* by 1200 AD. The artists in Japan were explicitly trying to capture the variational quality of natural forms.

Figure 61: Suminagashi (literally 'floating ink') was used to create natural forms.

Generative art created by computers can share this property of serendipity, of finding beauty in forms that resemble nature. Choosing which generative art to display is something like the role of a rock collector, or a child pressing fall leaves. The forms developed on their own, and the role of the collector is to pass aesthetic judgment on them.

Figure 62: Raku pottery from the Raku Museum

Figure 63: L-system based bioforms by William Latham

In my opinion, the most successful generative art tends to use colors in a way that emphasizes the relationship with nature, rather than using the full spectrum of saturated color that is the usual first approach as artists begin to explore this field This suggests a direction for future generative art. A machine learning algorithm could be trained to favor nature scenes, by presenting it with many images downloaded from the web, and giving a positive score to those tagged with natural terms such as "leaf," "landscape," or "canyon." Then some kind of completely separate fractal generative process could be set in motion, choosing its colors and other parameters randomly. These images would then be evaluated by the trained module, and those that most resembled the natural images it had seen would be kept, while the others would be discarded. Whether or not this additional step would result in aesthetically superior output would depend on how well the learning captured the relevant properties of the nature images.

Fiction and Reality in Photography

Early photographers attempted to imitate the artistic styles popular at the time. It was perhaps inevitable, given exposure times of up to an hour, that early photographs would be carefully staged. Yet as the process improved, these "tableau vivant" images became marginalized, and the authenticity of a photograph became a key element of how it is judged. Although the photographer may in fact spend a great deal of effort setting up a photograph, we require the illusion of spontaneity and realism. Contrast this with film; with film we expect fiction, we think nothing of an elaborate set, of actors portraying other cultures, or building up a drama. There are a few artists who create deliberate fiction with their photographs, but it is certainly the exception rather than the norm. Even in photographic portraiture, which is usually carefully arranged, the main criteria by which it is judged is whether it has captured the truth of the subject.

This attitude towards photography celebrates the individual and the particular, but some early photographers were more interested in discovering what was general. Could photography somehow be used to detect universal traits? Phrenology was a popular technique in the early 1800s, and while it has since been proven false, there is nothing inherently absurd in the possibility that particular character traits correspond with certain skull shapes. Unfortunately, such research wasn't carried on scientifically, but rather was used as an excuse to perpetuate racism and eugenics. Using multiple exposures, Darwin's cousin Francis Galton created examples of stereotypical head shapes for various traits. He noticed a phenomenon that still affects face research: the blended faces lost their individual distinctions, and by becoming generic achieved a kind of beauty that can also be created with makeup or a soft focus. By aligning facial features, a kind of early morphing was possible.

Morphing

With the low resolution of these photographs and the similarity of most facial proportions, an approximate alignment was acceptable, and this "cross-fade" was a popular technique in special effects prior to the use of computer graphics for morphing in *Willow* and *Terminator II*. For true morphing a mapping needs to be established between key points in the original and the final image. The first person to really explore such mappings was the naturalist D'Arcy Thompson, author of the classic book *On Growth and Form*. He noticed that many classes of living things (such as mammals, fish, or flowers considered as a class) shared common parts, which could be made to correspond by appropriate stretching and shrinking. This work has actually been quite useful in the development of image processing and image recognition techniques. (Coincidentally, the scan of this very page I had was warped from the curvature of the book pages, and I corrected the warp in Photoshop using a "distort transform" which overlays a similar grid and allows the user to move the grid points independently, warping the underlying image.)

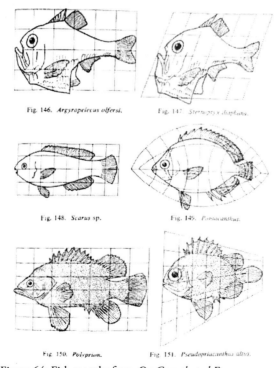

Figure 64: Fish morphs from *On Growth and Form*

PORTRAITS.

Groups I. II. III. and IV. V. VI. respectively illustrate a type of features common among men convicted of crimes of violence.

I II III

IV V VI

COMBINATIONS of PORTRAITS.

The Portraits of many different persons who have the same general type of features are here combined into single figures.

I, II and III

IV, V and VI.

FOUR PERSONS
None of the above Six

SEVEN PERSONS
including I, II and III

EIGHT PERSONS
including IV, V, and VI.

Photography as Memory

Samuel Butler (see chapter 9) was an enthusiastic amateur photographer. His ideas about memory were clearly shaped by his familiarity with the photographic process:

> Memory is a kind of way (or weight—whichever it should be) that the mind has got upon it, in virtue of which the sensation excited endures a little longer than the cause which excited it. There is thus induced a state of things in which mental images, and even physical sensations (if there can be such a thing as a physical sensation) exist by virtue of association, though the conditions which originally called them into existence no longer continue.
>
> This is as the echo continuing to reverberate after the sound has ceased.[5]

The nature of photography allows it to record details that no artist thought were significant at the time. There were a few precursors to photography that created prints with the same property of inadvertent detail. The earliest of these were the prints of hands, birds, and other small items created by spraying pigment at a stencil in front of a cave wall. From these, anthropologists can make educated guesses about the ages and genders of the artists.

Later printing techniques pressed plants flat, either with ink directly applied or into a thin metal sheet to create a mold that was used for printing. Photography can be seen as a continuation of this tradition of making prints from nature.

This connection with memory has always been an important aspect of creativity. The *Ars Memoria* was not just a way to commit things to rote memory, but a method for structuring that memory so that it would become a usable machine, a *machina mentis,* that would be able to work with the elements held in memory to create new ideas. As we develop artificial associative memories, they may prove useful to creative engines as well.

This chapter has looked at artistic machines as they are used by artists. All artists rely to some extent on fortunate accidents in their

---

[5]Samuel Butler, *The Note-Books of Samuel Butler*, p. 58

work, but some seek to emphasize this aspect of the process. With photography, the role of the artist as one who chooses to present an image as art is brought to the fore. Many artists using generative techniques rely on the kaleidoscope technique, and ultimately face the same issues of diminishing interest in individual pieces as the underlying pattern becomes clear. The final chapter will take a look at some possible ways to get around these limitations. In particular, the processes of reinterpreting accidents in a new way and selecting works that conform to a standard of interest or beauty are two processes that could in principle be automated.

XII
The Beginning

We have not yet created a machine that is widely accepted as being truly creative. The present is still, like the examples examined in this book, part of the *pre*history of creative machines. Consider the thread of influence that begins with Arabic divination devices which influenced Raymond Llull, who in turn influenced Leibniz, who influenced Babbage, who influenced Turing and the pioneers of electronic computing. In each step the scientist or philosopher makes clear in his writings that he was inspired to think about the problem by the earlier work in trying to build a mechanical mind. Their practical solutions weren't the critical thing; it was the idea that lived on.

In the following table some of the most significant examples of generative machines that were actually constructed (as opposed to being merely imagined) have been gathered.

Inventors and artists have been working on machine creativity for a long time. We know one way to go about it, the way used in most of the machines listed in the table below. One selects certain elements with an inherent beauty or meaning, invents rules for recombining them in an orderly way, and then randomly chooses new combinations. Most modern attempts at generative art can be seen as more complex versions of this same basic pattern.

Such machines are capable of generating variation, but sooner or later all the output from any one of these processes exhausts the space of possibilities inherent in its construction. There may be ways to move beyond this, to create machines that will continue to grow on their own and surprise us in ever new ways. If we leave aside the criteria that require conditions on the artist's intention, and judge the artwork by its own merits, the art from such a machine would be judged to be creative, relevant, beautiful and interesting.

Machine	Source of randomness	Pregenerated elements	Source of order
Divination machines	Dice, scattered seeds, etc...	Meaning of individual symbols	Rules of divination process
The *Ars Magna* of Ramon Llull		Meaning of initial terms	Rules of interpretation
Logic machines		Meaning of initial terms	Rules of inference
Eureka	Rotating drum	Latin terms	Rules of grammar
Kaleidoscope	Shaking	Bits of glass, etc...	Mirrors at rational angles
Harmonograph	Chaotic dynamics		Harmonic motion
Photography	Accidental effects	Image of scene	
Raku pottery	Accidental effects	Shape of pot, colors of glaze	Potter's wheel
Suminagashi	Fluid flow	Colors of ink	brushstrokes
Aeolian harp, wind chimes, etc...	Wind and chaotic dynamics	Tone of individual chimes	Harmonics between chimes
Dice music and the *Arca Musurgica*	Dice	Precomposed measures	Rules of combination
Componium	Loosely spinning pulley	Precomposed measures	Implicit rules in design of barrels

General knowledge

It is impossible for a system limited to a single narrow field of interest to apply new ideas from other fields. Generative art programs to date have a very limited world to work within. Attempts to include the larger world have mainly focused on bringing in content without any attempt at comprehension. As an early example, Brewster built kaleidoscopes that were open-ended: they could be pointed anywhere in the world and impose a sixfold symmetry on the viewpoint. Newer systems may download text or images from the web based on keyword searches, but such a system still only treats images as patches of color. Until information can be put into context, the artwork it produces will be like a collage made by a blind artist.

The Semantic Web project started by Tim Berners-Lee can be seen as a large-scale attempt to feed massive amounts of data into a traditional top-down AI. Google as a company is very aware of their position as a provider of practical bottom-up artificial intelligence. The combination of these efforts, and similar work by other companies and individuals, will form a large base of common knowledge from which to begin. A system based in such work could realistically be expected to be able to analyze the text (using natural language grammars) and generate appropriate illustrations (using portions of images labeled for web pages) for a book for young children.

Theories of how the world comes to be understood

Take music as an example. We are only beginning to have an understanding of the exact principles underlying the composition of melody. Before computers can create art, we must turn that art into a science. The field of computer vision is gradually making progress into such issues as recognizing objects and human figures, reconstructing a 3D scene from 2D images, understanding the general subject of an image, following subjects through the frames of a movie, finding salient parts of the image, and so forth. Once a system that can interpret images has been built, it can be incorporated into a feedback loop in the creation of visual artwork. Pieces generated by some kaleidoscopic code can then be evaluated by a separate vision system, looking for pieces that have greater human meaning, greater visual interest, and greater naturalism. Only those pieces that have passed this evaluation would then be presented as the creations of the computer artist, echoing the self selection that happens before an artist ever puts pen to paper.

The combination of greater knowledge of wide areas of the world and ways of perceiving what they have created that are separate from the means used to create them will lead to unanticipated creation. When AARON, Harold Cohen's artist program, sets out to draw a human figure, it always follows the same pattern (though with random parameters, such as the number, location, and pose of figures and background objects). But imagine that some future improvement of AARON is able to recognize that in one particular drawing a line drawn, intended to be a nose, had accidentally extended up to make an enquiring eyebrow. The program might be capable of recognizing that the resulting face held more visual interest because it had more expressive character. In that case, we might feel freer to acknowledge that by choosing to keep this particular drawing and exhibit it for our evaluation, the machine was showing a deeper level of creativity.

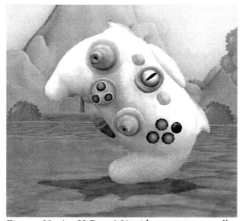

Figure 65: An X-Box 360 video-game controller created using the Spore creature creator.

This *misuse* of parametric invention is frequently seen in creative work. One popular creative digital tool in 2008 was the creature designer for the video game SPORE. All the possible "life forms" that could be created by this system are limited to the initial forms chosen by the game designers and the parametric sliders which define their size and placement. The designers intended for all the creations to have a kind of bubbly, cartoony character, and chose forms accordingly. Some people using these tools, however, found that it was possible to use elements in unusual ways: making teeth from elements intended as scales or feathers, making eyes from elements intended to be bony plates. With this expanded palette, they were able to create creatures that the authors of the game had never envisioned. From what was intended to be a single creature, they cleverly made what seemed to be a pair of acrobats grappling, or a fat man sitting on a walking chair, or a video game controller. This was possible because they perceived the results of their creation in a different way than the game's internal representation. What it represented internally as legs of a creature the artist and the audience were able to perceive as legs of a chair.

We can also sometimes see this in natural evolution, where, for example, the abdomen of an ant is reshaped by natural selection to resemble a spider, since this prevented the ant's ancestors from being eaten quite so often. Evolution has acted in a creative way by making a kind of spider costume for the ant.

A robust theory of beauty

The idea of a "theory of beauty" is perhaps as much an oxymoron as "creative machine." Ramon Llull[6] developed a wide-ranging theory of beauty in the 13th century as part of his great work, and the field of aesthetics has been explored in an idiosyncratic way by philosophers and critics. While art criticism is a rich field of insights, it is criticism written for other people, who already share a wide cultural vocabulary and instinctive reactions. What computers need is a theory of art criticism that could be communicated to a completely alien species, who don't even share the same senses as humans.

At earlier times in history, this would have been considerably easier. Codifying the ancient Egyptian theory of aesthetics in architecture, or that of the Roman Empire, for example, into a generative grammar would have been fairly straightforward. In the twentieth century, however, the very idea of aesthetics guiding artistic creation fell under attack. What are left are the criteria of novelty, significance, and inte-

---

[6] See Chapter VI for more about Ramon Llull.

grity. These are much harder to define, even in a way that people can agree on, let alone machines. In fact, art may have moved in this direction precisely because of the ability of machines to reproduce artwork exactly or with variations. In many ways, the contemporary art community isn't looking for beauty at all. Perhaps we should be trying to find ways to be creatively shocking, disturbing, unusual or cryptic instead.

Figure 66: The most popular nature and animal photography from the website DeviantArt. Images tend to be colorful, safe, awe-inspiring or cute, the kind of thing one would expect to see on a poster. There are four kittens, five trees, two flowers, three calm seas, and four images with distance fog. It seems likely that much of what makes these images have mass appeal could be learned by an image recognition algorithm. However, many of the most popular images in other categories are humorous images, whose humor seems fairly difficult to capture with any AI system we can build in the foreseeable future. (browse.deviantart.com/photography/nature/)

On the other hand, if we leave aside the contemporary art world and look instead at the art the majority of people actually create and consume, finding guiding aesthetic principles is less difficult. One can explore this popular aesthetic by looking at visual art marked as "popular" on sites such as deviantart.com or flickr.com. These images tend to be colorful, emotionally immediate (sentimental, shocking, or humorous), visually arresting and clearly defined in details and subject matter. They are "eye candy," meaning that like candy they directly and unsubtly play on the senses in a way that gives an immediate positive emotional response. Although the cultural styles have changed, many of these aesthetic attributes seem likely to be human universals that might be learned. These features can be studied and quantified by researchers.

Even so, a theory of beauty is so contradictory, so self-defeating, that it seems to only really occur to those with an unusual way of thinking to attempt it. The troubled semi-autobiographical narrator Phaedrus of *Zen and the Art of Motorcycle Maintenance* is one example. Another example is the eccentric scientist Patrick Gunkel. His theory of Ideonomy is a fascinating mix of clever ideas, maps of the unmappable, and sheer crackpottery. Among other projects, he attempted to map out the space of human conception of beauty. To do this, he found examples of what people found beautiful and then considered how similar each was to the others. This data was then fed to a computer program that used multi-dimensional scaling to transform the results into a two dimensional map, where ideas that were similar to each other were plotted close together, and those that were different were plotted far apart.

Here is his list of the types of beauty he considered (his ordering):

Academic graduation ceremony
Acquisition of first home
Apocalyptic psychostasia
Athletic competition

Attractive voice

Aurora
Beatitude (consummate bliss)
Bees' pollinating flowers
Biological evolution of ontogeny
Birdsong
Birth of one's child

Bold architecture or elegant bridge

Brilliant attire
Bubble bath
Butterfly wing color patterns

Wonderfully fortuitous event
Wedding
Waterfall, river rapids, or freshet
Vista from mountain top or canyon rim
Visionary statesmanship or sea-green incorruptible
Victory in war
Unity of natural laws
Unflagging loyalty of spouse
Twilight mystery
Tragic love
Tiny humans at foot of mountain
Time-lapse film of sky, anthesis, or child's growth
Supreme amical moment or act
Sudden unexpected resolution of crisis
Sublime wickedness

Calligraphy	Sublime dream
Cancer patient's indomitable will to live	Springtime
Catharsis	Spaceship launch
Cavern decorated with speleothems	Snowfall
Charitable act or self-giving	Skyful of soaring thunderheads
Charmed banter, dalliance, or dazzling laughter	Skydiving or gliding
Child's toys	Singular humility or simplicity
Coitus	Sights and aroma of meal being prepared
Concept of infinity or Apeiron	Seashore (surf, immense beach, or towering cliff)
Cooperative endeavor	Saintly kindness
Coral reef	Rose
Coruscant stars across night sky	Revelatory insight or epiphany
Cosmos-atom size ratio (mental juxtaposition)	Reunion with childhood friend
Jewel	Industrious ant colony
Jupiter's surface or Saturn's ring system	Individual or panhuman wisdom
Justice, or villainy receiving its due	Impressionist painting
Lavish banquet	Hydrologic cycle
Lush and surreal rain forest	Human progress
Machinery of the mind and brain	Human ideals
Magnificent battleship	Harp's sound
Magnificent body (figure or physique)	Handsome face
Marine islet or archipelago	Glorious pageantry or pomp
Massive or lofty tree	Glamorous scientific laboratory
Maternal or filial devotion	Geometric proof
Melody, song, or chorus	General prosperity
Meteor shower or radiant, or spectacular comet	Forest fire
Microscope, telescope, or other instruments	Foreign travel
Old-age reminiscence and reverie	Exuberant elfin child
Palatial estate	Extraordinary wealth or good fortune
Patterns of frost or dew	Epic interpersonal tableau (eg diplomatic)

Personal grace or magnanimity	Elevatory metanoia[7]
Probity and zeal for truth	Drifting bubble
Profound metaphor	Dramatic chess game, position, or tactic
Rainbow or green flash	Destiny or time's river
Rescue from misery or horror	Democratic processes

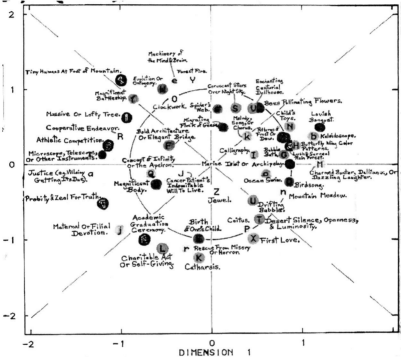

Figure 67: Map of types of beauty from Patrick Gunkel's Ideonomy

One can perhaps make out features such as classical beauty on the left side of the plot and romantic beauty on the right, but it's unclear how to make use of such a graph. The list does give a sense of how broad a topic would need to be addressed by a serious attempt to teach beauty to a machine.

---

[7] Perhaps only Gunkel himself could tell us exactly what he meant by "Elevatory metanoia," "Supreme amical moment," or "Apocalyptic psychostasia."

It seems unlikely that one could spell out explicitly the limits of what we as humans find pleasing: the field is too large, diverse, and amorphous. The only approach with a hope of success is some kind of training done on an enormous number of examples. On the other hand, it may not be necessary for a machine to hold a model of everything we hold to be beautiful; a small corner of the space may be enough to produce beautiful and novel results.

Information theory

Information theory is a way of measuring the information content in a string of symbols. Consider the message

AAAAAAAAAAAAAAAAAAAAAAAAAAAAAAAAAAAAAAAA[8]

This message doesn't contain much information. We can write a short program to get the same results:

```
Print A 40 times.
```

Some messages are compressible like this: we can write a shorter message that gets exactly the same message across. Consider this one:

```
THS MSSG HS BN CMPRSSD
```

We are still able to read it, hinting that the vowels contained redundant information. Of course, these two examples used your brain to do the decompression, but it isn't hard to program a computer to do the same kind of elimination or restoration of redundant information. ZIP files work this way. In *algorithmic information theory*, the information content of a message is the length of the shortest program (in a particular programming language) that will generate the message.

One way of looking at this is as a kind of automation of science or learning. In science, we take data and try to come up with a short formula to explain the data. The shorter the formula and the less data left unexplained, the more elegant the theory.

What's interesting about all this, from an artificial creativity standpoint, is that a decompression algorithm is a way of creating human understandable text from a much shorter input. In fact, a

---

[8] *Monty Python and the Holy Grail.* He must have died while writing it.

perfect decompression algorithm would take random input and still output perfectly readable text. This was noticed by Claude Shannon in his 1948 paper defining information theory.

In the first example, he chooses letters randomly:

```
XFOML RXKHRJFFJUJ ZLPWCFWKCYJ
FFJEYVKCQSGHYD QPAAMKBZAACIBZLHJQD. 9
```

It is just a mess. In the second example, he chooses letters as frequently as they appear in real English sentences:

```
OCRO HLI RGWR NMIELWIS EU LL NBNESEBYA TH
EEI ALHENHTTPA OOBTTVA NAH BRL.
```

It at least has separate "words," because spaces appear more frequently than once every 26 letters on average, but still doesn't have much structure.

Then he chooses letters based on the previous letter:

```
ON IE ANTSOUTINYS ARE T INCTOREST BE S
DEAMY ACHIN D ILONASIVE TUCOOWE AT TEASO-
NARE FUSO TIZIN ANDY TOBE SEACE CTISBE.
```

For example, N frequently follows O in English, so after choosing O randomly the next letter chosen happened to be N. It could have been F (another letter that frequently follows O) but the chance of picking Q was very close to zero. A few other short English words are also present, and many of the other words, such as INCTOREST and DEAMY, are at least pronounceable. Even such a simple model has managed to capture something of the structure of English.

By looking at the previous *two* letters, it gets still more like English.

```
IN NO IST LAT WHEY CRATICT FROURE BIRS
GROCID PONDENOME OF DEMONSTURES OF THE
REPTAGIN IS REGOACTIONA OF CRE.
```

Now, imagine sitting in the passenger seat next to someone who is notoriously bad at remembering directions. The directions for driving are very complex, but it's unnecessary to tell the driver to stop at every stop sign, or keep going straight past every side street. The directions can be compressed by only giving directions when the driver seems about to make a mistake. In the same way, a Markov text generation

---

9 Claude Shannon, "A Mathematical Theory of Communication," *Bell Systems Technical Journal* Vol. 27, 1948

mechanism based on letter frequencies that has already generated the letters INTERESTI doesn't need any help to come up with NG for the next two letters. The letter after that is probably a space, but might be an N (to spell INTERESTINGNESS) or an L (to spell INTE-RESTINGLY) and is almost certainly not going to be any other letter. Which one it is can be communicated with only a couple of bits of information. For this reason, the better one can anticipate a text, the more it can be compressed.

Now consider the same process, but instead of using letters, we look at the frequency of words. This looks at just the previous word:

```
REPRESENTING AND SPEEDILY IS AN GOOD APT
OR COME CAN DIFFERENT NATURAL HERE HE THE
A IN CAME THE TOOF TO EXPERT GRAY COME TO
FURNISHES THE LINE MESSAGE HAD BE THESE.
```

This looks at the previous two words:

```
THE HEAD AND IN FRONTAL ATTACK ON AN ENG-
LISH WRITER THAT THE CHARACTER OF THIS
POINT IS THEREFORE ANOTHER METHOD FOR THE
LETTERS THAT THE TIME OF WHO EVER TOLD
THE PROBLEM FOR AN UNEXPECTED.
```

It is almost comprehensible text! Google has made publicly available the statistical tables needed to continue this up to looking at the previous *four* words, based on statistically analyzing a good fraction of all the text on the Web. By choosing a fifth word based on looking back at the previous four words, almost all sentences generated are grammatically correct and comprehensible.

While this works to generate plausible phrases and sentences, whole paragraphs or pages of this are still nonsense: they seem to mean something for a moment or two, but like the ramblings of a madman, they don't go anywhere. The decompression still is just based on individual words, not taking into account the meaning that comes from considering the larger context. Algorithms have recently been developed that use semantic information to further compress text. For example, if a paragraph is talking about turtles, the word "shell" is more likely to come up than in normal text, so it can be assigned a shorter code word to be used in the compressed paragraph. Text generated randomly by such an algorithm would seem to stick to certain themes, as if it is talking about an idea and developing the idea through a paragraph.

This "semantic" information doesn't come from sophisticated natural language parsers. Instead, it just looks at the surrounding words in a paragraph, and assumes if two paragraphs have a similar distribution of words, they mean about the same thing. Obviously this isn't strictly true: "I ate the chicken and the corn" doesn't mean the same thing as "The chicken and I ate the corn." But in practice, it gives amazingly good results. For example, on an SAT test of synonyms or word analogies, it performs better than the average high school student. There are some AI researchers, such as Douglas Hofstadter, who believe that analogy-making is the key ingredient in creative thought.

Image Analogies

Some of my own research in the area of artificial creativity has been building on the "Image Analogies" research project of Aaron Hertzmann. Analogies are of the form

cat : meow :: dog : bark.

This means "The relationship of the word *cat* to the word *meow* is the same as the relationship of the word *dog* to the word *bark*." With image analogies, each of these words is replaced by an image. The first image might be a photograph of a cat, and the second image an oil painting of the same cat from the same point of view. We then supply a photograph of a dog, and the task of the program is to create the image of an oil painting of the dog. The analogy in this case would be

photo of cat : oil painting of cat :: photo of dog : _____

where the blank is filled in with "**oil painting of dog**." The program looks at small bits of the cat photograph—the corner of an eye, the tip of the nose—and tries to find matching bits in the dog photograph. When it finds them it copies the corresponding bit of the cat oil painting, and pastes it in to form a piece of the dog oil painting. (Care is taken with the edges of the bits to do this in such a way that they blend together, hiding the fact that the image is really a patchwork.)

This works well as long as the correspondence between the photograph and the oil painting of the cat is direct. As soon as the artist starts taking liberties with the proportions or placement of the cat within the scene, so that the details of the two images no longer line up, the process fails to work.

The "oil paintings" generated by such a system have perspective, correct proportions, fine brushwork—all marks of a skillful artist. But they don't come from skill on the part of the machine, any more than the perspective in a photograph comes from skill on the part of the camera. Notice how different this is from the way a child learns to draw. Children start somewhere else, with images that are more like writing than like photography. The images stand for the things they are meant to represent. A circle stands for a body, two straight lines stand for legs, two dots stand for eyes, and the whole thing stands for a person. What the child is drawing isn't so much what is seen, but some kind of internal model of a person. It takes a great deal of training and effort to get beyond this sign-based artwork.

A machine able to do the same trick that a three-year-old can—look at a picture of a cat, recognize it and represent it by a sign for cat—would be a step closer to being a truly creative machine. The analogies discovered by such a machine wouldn't be brittle surface analogies, but deep conceptual ones.

A Unified System

While all of the ideas mentioned in this chapter would make generative art richer and more meaningful, it is not at all clear that this would lead to a machine that would be able to exhibit creativity. Imagine what a system that incorporated all of these suggestions would be like.

In our imaginary program, a user would choose a general theme, like "seascape." This is the limit of the user's interaction with the machine. The program uses an image search engine to find thousands of images of seascapes. Using image recognition technology, it identifies the regions of many of the images as containing sea, sky, beach, and rocky forms. From these images it constructs a generative grammar of generic seascape images.

Figure 68: An example of Image Analogies from the project page at NYU. Upper left: photograph of a swan. Upper right: Pastel sketch of a swan drawn by a human artist. Lower left: photograph of a woodland scene. Lower right: computer generated "pastel sketch" of the woodland scene. The style of the strokes is copied from the swan sketch above.

Using its knowledge of human aesthetic preferences, it chooses to create a scene near sunset, when the colors are most saturated. (On another run, it might randomly decide on pastel colors, or a different arrangement of colors that it recognizes as fitting into some human pattern of color harmony.)

It then explores the space of words associated with sea, beach, and sky, common nouns that tend to appear together in texts with these words. It comes up with the idea of a shipwreck, and adds that to the scene, expecting it to increase human interest. It searches the internet for images of shipwrecks, and finds one that satisfies the parameters it is searching for (the image is large enough, shows the shipwreck against an easily removed background, has shading that enables it to estimate 3D shape more easily, and so forth.) It builds an estimated 3D model from the image and rotates the model to find a new perspective.

At this point it only has a plan. It chooses a rough layout for what regions of the image should be sea, sky, rocks, beach, and shipwreck, and renders these using image-analogy type techniques to patch together small details from all the images it has seen. It renders several different versions. It then performs image recognition again, on its own constructions. It evaluates these against several different measures of visual interest (based on Shannon entropy, for example), beauty (looking for curves humans are known to find pleasing, color combinations known to be harmonious and creating a mood consistent with the emotional connotations typically associated with seascapes and shipwrecks), and realism (looking for physically impossible anomalies or rendering artifacts).

If none of the images are satisfactory, it might try again, backtracking one step to generate more images, or going back more steps, perhaps throwing away the idea of a shipwreck altogether. Or, if the result passes these tests, it could go on to "paint" the image, using a physical model of paint, canvas, and brush to render the image from its "mind's eye." Again, it could render several versions using different brush parameters and palettes and choose the one that best fits its model of "pleasing."

In addition to all this, the program would be able to modify any step of this process on rare occasions, deliberately choosing to do things that are against its own artistic conventions. Someone looking for a more avant-garde robotic artist could turn up this random self-modification parameter.

What has just been described is a little beyond what can currently be built, but it doesn't seem like any of it is unachievable. Many of the pieces described are active areas of research by hundreds of scientists. A rudimentary version of all of this could be built today, and improvements could be put in place in a piecewise fashion as advancements are made.

A more serious objection is that such a system still would not be creative. Such a system would surely fail to live up to the standards we have for professional visual artists—the output would still fall into the category of kitsch. It would be able to "learn" by taking in new imagery from the web, but in terms of its own process, no growth would really occur. Self-modifying code is possible, and the results can be unpredictable and fascinating, but a key problem is making sure that modification is an improvement. An artificial device doesn't have any concept of "better" artwork except what we give it. Yet a plausible definition of creativity is finding a *new* way to improve.

Those who feel that this means the system is not "creative" probably will need to wait a lot longer to be satisfied. There are many problems that computer scientists informally classify as "AI-Complete." These problems we don't realistically expect to be solved until a program is written that can act indistinguishably from a human, able to pass the Turing test, and any reasonable variation on it, fairly easily. It may be that true creativity is that hard.

For others, even that will not suffice, since there is no evidence that the machine "artist" is feeling anything (in the sense of qualia) as it produces the artwork. These critics would say that its desires are not real desires, its choices are not real choices, and its art cannot properly be called that at all. Any credit must go to the humans who designed these systems. For these critics, the nature of the production of the piece always supersedes any aesthetic qualities the finished piece may have.

Making peace

There has been a continuous trend in society to redefine art to its aspects which cannot be automated. In the original definition of the word, there was no distinction made between simply creating a copy and creating a something unique: both were considered artwork. However, as the process of creating copies became automated, originality was seen as the key thing in a work of art. As long as copies were technically inferior, craftsmanship was prized; but when a photograph passed the realism that a painter could capture, accuracy was no longer seen as the primary virtue, and new styles of art that emphasized experimentation and originality became more valued. When player pianos became popular, the ability to play strings of notes quickly (known as virtuoso playing) was devalued, and the things a player piano couldn't reproduce, such as variations in volume and expression, were seen as the most important thing for a pianist to master.

The attempt to understand creativity and build machines that can make beautiful things with less and less guidance from us is not any kind of a threat. Instead, like any new art, it is an opportunity to explore new ideas and to recognize previously invisible limitations of the old. Artists will work with and respond to these new tools.

In the modern world, machines have led to a lot of ugliness, the sameness of advertising and strip-malls, blocks on blocks of identical grey Soviet apartments. In the name of efficiency, we as a society have often sacrificed aesthetics and uniqueness. Exploring these ideas about the nature and meaning of art and creativity can help us to find ways to use machines to reclaim some of the lost beauty that used to fill the places that we live.

Bibliography

Works in the following list were used as research for this book. One particularly good reference for carrying the general subject of this book into the twentieth century is Margaret Boden's *Mind as Machine*. I have not listed the works of Athanasius Kircher or Ramon Llull, as few have been translated from Latin into English or reprinted, but they are worth searching for online, if only to look at the marvelous illustrations. Other books are referenced in the footnotes that I have not included here. Please let me know if there are ideas or images, from these books or others, that I have not given proper attribution.

For additional related material, please visit the website created for this book: **machinamenta.blogspot.com**.

Altick, Richard D., *The Shows of London*

Babbage, Charles, *Passages from the Life of a Philosopher*

Bailey, James, *After Thought: The Computer Challenge to Human Intelligence*

Batchen Geoffrey, *Burning with Desire: the Conception of Photography*

Bergson, Henri, Arthur Mitchell, *Creative Evolution*

Boden, Margaret, *Mind as Machine: A History of Cognitive Science*

Butler, Samuel, *Erewhon*

Cicero, Marcus Tullius, *The Treatises of M. T. Cicero: On the Nature of the Gods; on Divination*

Cook, Simon, 'Minds, Machines and Economic agents: Cambridge Receptions of Boole and Babbage,' *Studies In History and Philosophy of Science* Part A, Volume 36, Issue 2, June 2005, Pages 331-350

Davis, Martin D., *The Universal Computer: The Road from Leibniz to Turing*

Donald, Diana, *Endless Forms: Charles Darwin, Natural Science, and the Visual Arts*

Dyson, George B., *Darwin Among the Machines: The Evolution of Global Intelligence*

Eco, Umberto, The *Search for the Perfect Language*

Fanger, Claire, Richard Kieckhefer, Nicholas Watson, *Conjuring Spirits: Texts and Traditions of Medieval Ritual Magic*

Gribbin, John R., *The Fellowship: Gilbert, Bacon, Harvey, Wren, Newton, and the Story of a Scientific Revolution*

Hanafi, Zakiya, *The Monster in the Machine: Magic, Medicine, and the Marvelous in the Time of the Scientific Revolution*

Hankins, Thomas L., Robert J. Silverman, *Instruments and the Imagination*

Hayes, Mary Dolores, *Various Group Mind Theories: Viewed in the Light of Thomistic Principles*

Hoffmann , E. T. A., *Automata*

Holland, John H., Keith J. Holyoak, Richard E. Nisbett, *Induction: Processes of Inference, Learning, and Discovery*

Houdin, Robert, *Memoirs*

Husbands, Philip, Owen Holland, Michael Wheeler, *The Mechanical Mind in History*

Ifrah, Georges, *The Universal History of Computing: From the Abacus to the Quantum Computer*

Jevons, William Stanley, *The Principles of Science: a Treatise on Logic and Scientific Method*

Ord-Hume, W. J. G., *Barrel Organ*

Peek, Philip M., *African Divination Systems: Ways of Knowing*

Raffaelli, Tiziano, *Marshall's Evolutionary Economics*

Rhodes, Richard, *Visions Of Technology: A Century Of Vital Debate About Machines Systems And The Human World*

Rossi, Paolo, Stephen Clucas, *Logic and the Art of Memory: the Quest for a Universal Language*

Smee, Alfred, *The Process of Thought Adapted to Words and Language*

Standage, Tom, *The Turk: The Life and Times of the Famous Eighteenth-Century Chess-Playing Machine*

Standage, Tom, *The Victorian Internet*

Strickland, Lloyd, *The Shorter Leibniz Texts: A Collection of New Translations*

Swade, Doron, *The Difference Engine: Charles Babbage and the Quest to Build the First Computer*

Tiggelen, Philippe John Van, *Componium: the Mechanical Musical Improvisor*

Venn, John, *Symbolic Logic*

Wilkins, John, *The Mathematical and Philosophical Works of the Right Rev. John Wilkins, Late Lord Bishop of Chester*

Woolley, Benjamin, *The Bride of Science: Romance, Reason, and Byron's Daughter*

About the author: Douglas Summers Stay works for the Army Research Lab, developing robot vision algorithms. He did graduate studies in computer science at New York University and the University of Maryland, College Park. He lives with his wife and son in Frederick, Maryland. He really likes Legos.

CPSIA information can be obtained at www.ICGtesting.com
Printed in the USA
LVOW121709280212

270834LV00012B/30/P